SUSTAINABLE ENVIRONMENTAL ENGINEERING

Water Security, Energy Conservation, and Green Process

SUSTAINABLE ENVIRONMENTAL ENGINEERING

Water Security, Energy Conservation, and Green Process

Edited by

Sonia Khanna, PhD

Ann Rose Abraham, PhD

Kian Hariri Asli, PhD

A. K. Haghi, PhD

AAP | APPLE ACADEMIC PRESS

First edition published 2025

Apple Academic Press Inc.
1265 Goldenrod Circle, NE,
Palm Bay, FL 32905 USA

760 Laurentian Drive, Unit 19,
Burlington, ON L7N 0A4, CANADA

CRC Press
2385 NW Executive Center Drive,
Suite 320, Boca Raton FL 33431

4 Park Square, Milton Park,
Abingdon, Oxon, OX14 4RN UK

Library and Archives Canada Cataloguing in Publication

Title: Sustainable environmental engineering : water security, energy conservation, and green process / edited by Sonia Khanna, PhD, Ann Rose Abraham, PhD, Kian Hariri Asli, PhD, A.K. Haghi, PhD.
Names: Khanna, Sonia, editor | Abraham, Ann Rose, editor | Asli, Kian Hariri, editor. | Haghi, A. K., editor
Description: First edition. | Includes bibliographical references and index.
Identifiers: Canadiana (print) 20240444019 | Canadiana (ebook) 20240444035 | ISBN 9781774916902 (hardcover) | ISBN 9781774916919 (softcover) | ISBN 9781003506430 (ebook)
Subjects: LCSH: Water security. | LCSH: Energy conservation. | LCSH: Green technology.
Classification: LCC TD345 .S87 2025 | DDC 628.1—dc23

Library of Congress Cataloging-in-Publication Data

CIP data on file with US Library of Congress

ISBN: 978-1-77491-690-2 (hbk)
ISBN: 978-1-77491-691-9 (pbk)
ISBN: 978-1-00350-643-0 (ebk)

About the Editors

Sonia Khanna, PhD
Assistant Professor, Department of Chemistry and Biochemistry, Sharda School of Basic Sciences and Research, Sharda University, Greater Noida, Uttar Pradesh, India

Sonia Khanna, PhD, is currently working as an Assistant Professor in the Department of Chemistry and Biochemistry at Sharda University, Greater Noida, India. She has teaching and research experience of over 14 years, with expertise in the field of inorganic materials, coordination compounds, organic-inorganic hybrid materials, and biological activity of materials. She has published 18 articles and six book chapters and has several patents and copyrights to her credit. She has supervised many MSc and PhD students and is currently associated with many research projects. She has been associated with projects sponsored by the reputed organization Indian Space Research Organisation, India. She earned her PhD in Inorganic Chemistry at Guru Nanak Dev University, Amritsar, India, and is recipient of a gold medal for her MSc and has many merit certificates for best performance in chemistry. She has qualified in many national-level competitive exams (NET-JRF, GATE).

Ann Rose Abraham, PhD
Assistant Professor, Sacred Heart College (Autonomous), Thevara, Kochi, Ernakulam, Kerala, India

Ann Rose Abraham, PhD, is currently an Assistant Professor at the Department of Physics, Sacred Heart College (Autonomous), Thevara, Kochi, Kerala, India. She has expertise in the fields of condensed matter physics, nanomagnetism, multiferroics, polymeric nanocomposites, etc. She has research experience at various reputed national institutes, including Bose Institute, Kolkata, India; SAHA Institute of Nuclear Physics, Kolkata, India; and UGC-DAE CSR Center, Kolkata, India; and she has collaborated with various international laboratories. She is a recipient of a young researcher award in physics and best paper awards (2020, 2021). She served as Assistant

Professor and Examiner at the Department of Basic Sciences, Amal Jyothi College of Engineering, under APJ Abdul Kalam Technological University, Kerala, India. Dr. Abraham is a frequent speaker at national and international conferences. She has a good number of publications to her credit in many peer-reviewed high-impact journals of international repute. She has authored many book chapters and edited more than 10 books with Taylor and Francis, Elsevier, etc.

Kian Hariri Asli, PhD
Research Scholar, Department of Electronic Engineering, University of Rome "Tor Vergata," Rome, Italy

Kian Hariri Asli is a Research Scholar in the Department of Electronic Engineering, University of Rome "Tor Vergata", Rome, Italy. He has expertise in electronic and control engineering. He works toward the development of energy saving by networked sensors, fuzzy logic, artificial intelligence (AI) and Internet of Things (IoT) based on geospatial information systems (GIS).

A. K. Haghi, PhD
Research Associate, Department of Chemistry, University of Coimbra, Portugal; Professor Emeritus of Engineering Sciences and Senior Material Engineering Consultant, UK

A. K. Haghi, PhD, is a retired professor who has authored, co-authored, edited, or co-edited over 1,000 publications, including books, book chapters, and papers in refereed journals. These works have garnered over 3,800 citations and an h-index of 32, as per the Google Scholar database. Currently, he serves as a research associate at the University of Coimbra, Portugal. Professor Haghi has been the recipient of numerous grants, has provided consultancy services to various major corporations, and is a frequent speaker at national and international events. He is the Founder and former Editor-in-Chief of the *International Journal of Chemoinformatics and Chemical Engineering* and *Polymers Research Journal*. Additionally, Professor Haghi has served on the editorial boards of several international journals and as a member of the Canadian Research and Development Center of Sciences and Cultures. He has supervised multiple PhD and MSc theses at the University of Guilan (UG) and has co-supervised international doctoral projects. Professor Haghi holds a BSc in urban and environmental engineering from the University of North Carolina (USA) and two MSc degrees, one in mechanical engineering from North Carolina State University (USA) and another in applied mechanics, acoustics, and materials from the Université de Technologie de Compiègne (France). He obtained his PhD in engineering sciences from Université de Franche-Comté (France) and is an active reviewer for esteemed international journals.

Contents

Contributors

K. B. Akhila
Department of Chemistry, CMS College, Kottayam, Kerala, India

Kian Hariri Asli
Department of Electronic Engineering, University of Rome "Tor Vergata," Rome, Italy

S. Sajith Babu
Assistant Professor, Department of Physics, Catholicate College, Pathanamthitta, Kerala, India

Anjika Boora
Postgraduate Student, Department of Chemistry and Biochemistry, Sharda School of Basic Sciences and Research, Sharda University, Greater Noida, Uttar Pradesh, India

Cintil Jose Chirayil
Department of Chemistry, Newman College, Thodupuzha, Idukki, Kerala, India

S. L. Divya
Department of Computer Applications, SCMS School of Technology and Management, Muttom, Kochi, Ernakulam, Kerala, India

Rama Dubey
SPD Division, Defense Research Laboratory, DRDO Tezpur, Assam, India

Dhiraj Dutta
SPD Division, Defense Research Laboratory, DRDO Tezpur, Assam, India

Sneha Jipson
Department of Chemistry, Newman College, Thodupuzha, Idukki, Kerala, India

Annu Jorly
Department of Chemistry, Newman College, Thodupuzha, Idukki, Kerala, India

Sonia Khanna
Assistant Professor, Department of Chemistry and Biochemistry, Sharda University, Greater Noida, Uttar Pradesh, India

Anitta Mathew
Department of Chemistry, Pavanatma College, Murickassery, Idukki, Kerala, India

S. L. Navya
Department of Physics, Sacred Heart College (Autonomous), Thevara, Kochi, Ernakulam, Kerala, India

Rony Rajan Paul
Department of Chemistry, CMS College, Kottayam, Kerala, India

S. S. Praveen
Assistant Professor, Department of Physics, St. Thomas College, Kozhencherry, Pathanamthitta, Kerala, India

Narender Singh
PhD Research Scholar

Sabu Thomas
School of Chemical Sciences, Mahatma Gandhi University, Kottayam, Kerala, India

Sunsu Kurian Thottil
Department of Physics, St. Teresa's College (Autonomous), Ernakulam, Kerala, India

Abiya Varghese
Department of Chemistry, Newman College, Thodupuzha, Idukki, Kerala, India

Roli Verma
Department of Chemistry, JSS Academy of Technical Education, Noida, Uttar Pradesh, India

Aswathy Vijayan
Doctoral Student, Department of Physics, St. Thomas College, Kozhencherry, Pathanamthitta, Kerala, India

Preface

This new title is a reference book covering the most relevant important topics of our time, from the impact of climate change to urban environment sustainability, using empirical and theoretical insights. It also studies water security and the meaningful connectivity between water security and energy conservation. The book demonstrates that water security and energy systems cannot be separated from the challenges associated with a sustainable environment.

The volume reports on the relationship between climate change and environmental sustainability with an emphasis on comparative case studies. It provides a broad overview of the interconnection between sustainability alongside population growth, rising living standards, and increasing demand for the consumption of water and energy. It draws the attention of readers to the interlinkage of climate change and the two sectors of water and energy that are very essential for attaining sustainable environmental goals.

This reference book explains what sustainable development is and how to achieve it. It shows that environmental sustainability goals can be achieved through the development of green materials. For instance, green functionalized materials are used in energy science and pollution management for future technology development and can establish a sustainable society in harmony with nature. The volume also reviews the growing threat of climate change that has a significant impact on water availability and energy demand.

This research-oriented book is a comprehensive guide that gives an overview of recent progress in sustainable developments related to the environment and its prospects. This volume is highly recommended for research communities, environmentalists, academicians, and scientists.

Key Features:

- Provides the most recent information about green process applications in a sustainable environment.
- Shows a wide range of connectivity between water security and energy conservation for environmental sustainability.

- Focuses on the utilization of water resources and energy conservation for sustainable development.
- Explores case studies in energy science and water pollution.
- Provides scientific information describing various examples that help gain knowledge for postgraduate research students.

CHAPTER 1

Fuzzy Logic for Smart Water and Wastewater Systems

KIAN HARIRI ASLI

Department of Electronic Engineering, University of Rome "Tor Vergata," Rome, Italy

ABSTRACT

In the intelligentization of water and wastewater facilities, the location of the cities and demographic changes in the future are important factors. The examination of the city texture, residential areas, commercial areas, and population density in the present and future according to the geospatial information system (GIS) map reveals the water and wastewater volume in the past years and its forecast in the future. In this work, the identification of the critical zones with the most need for drinking water was carried out by fuzzy logic, remote sensing (RS), and the internet of things (IoT). This work showed that networked sensors, GIS, RS, and IoT can make predictions of water resource changes in a short time. The results showed that a decrease in water working pressure reduced water leakage. This work showed that dynamic and moving installations can be guided and controlled in the best way with the help of fuzzy logic.

1.1 INTRODUCTION

Today, the issue of water supply is one of the most important concerns of mankind. For this reason, water supply is one of the most important concerns of governments now and in the future, and therefore large investments have been made in this sector.

Sustainable Environmental Engineering: Water Security, Energy Conservation, and Green Process.
Sonia Khanna, Ann Rose Abraham, Kian Hariri Asli, & A. K. Haghi (Eds.)

According to an estimate, one-third of the world's population is under water stress and this threat may become more serious in the coming years.

The purpose of the intelligentization of water purification, storage, transmission, and distribution facilities and wastewater collection, treatment, and disposal facilities is technical, economic, and environmental improvement and ultimately saving water and energy consumption.

1.1.1 FUZZY LOGIC AND SMART SYSTEMS

The fuzzy logic method and data intercommunication are complex mathematical algorithms in the context of GIS affected by internal and external factors in smart systems. Some variables are of two types in calculations: quantitative values that can be expressed with a certain number and qualitative values that are expressed based on a characteristic. These two values are convertible. But in the fuzzy set, each of these attributes is defined based on the membership function and valued between zero and one. Therefore, for the rapid data intercommunication in all kinds of complications in water systems, it must be defined the classes and sub-classes related to water supply, storage, transmission, and distribution facilities. Hence, it is necessary to create and innovate new and multi-valued logic related to fuzzy logic [1–10].

Fuzzy logic is a new technology that replaces methods that require complex and advanced mathematics to design a model for the system by using linguistic values and expert knowledge. The human brain defines and values sentences by considering various factors based on inferential thinking. Fuzzy logic modeling based on the human brain in language and mathematical formulas will be a very complicated task. Fuzzy logic is one of the multi-valued logics and relies on the theory of fuzzy sets. Fuzzy sets are derived from the generalization and expansion of definite sets in a natural way. Crisp sets are the normal sets that are introduced at the beginning of the classical theory of sets. Adding the definite attribute creates a distinction that helps one of the innovative and vital concepts in fuzzy logic known as the membership function to easily come to mind. In the case of deterministic sets, the membership function has only two values. In its range (in mathematics, the range of a function is equal to the set of all outputs of the function), yes and no (one and zero), which are the same two possible values in classic two-valued logic.

There is the membership function of the element in the definite set. The range of the membership function in the case of deterministic sets becomes a closed interval for fuzzy sets.

For example, water pressure can be considered a linguistic variable, which can contain values such as low, high, weak, medium, and strong. In mathematical language we have (P = Pressure):

$$P\ (\text{pressure}) = \{\text{low, high, weak, medium, strong}\} \quad (1)$$

The degree of membership indicates the degree of membership of the element to the fuzzy set. If the membership degree of an element of the set is zero, that member is completely out of the set, and if the membership degree of a member is equal to one, that member is completely in the set. Now, if the degree of membership of a member is between zero and one, this number represents the degree of gradual membership [11–14].

1.1.2 FUZZY LOGIC MATRIX AND QUANTITATIVE STRATEGIC PLANNING

A fuzzy logic matrix and quantitative strategic planning are analytical methods to determine the relative attractiveness of strategies. By using this method, it is possible to objectively determine various strategies that are among the best. To prepare the desired matrix, the following steps must be taken:

- Insert major foreign opportunities and threats, internal strengths, and weaknesses in the right column of the matrix.
- Insert coefficients of internal and external evaluation in the matrix.
- Determine selected strategies: 1 = not attractive, 2 = somewhat attractive, 3 = reasonable attractiveness, and 4 = very attractive.
- From the mathematics of fuzzy logic and arithmetic mean, the highest score indicates the more attractiveness of the strategies.

Fuzzy logic and SWOT method for water and wastewater projects SWOT (strengths, weaknesses, opportunities, and threats) in compliance with the Fuzzy logic method can be used to identify strengths, weaknesses, opportunities, and threats for water and wastewater projects. The final phase of the strategic environment research project using the SWOT method gives a coefficient and rank to each of the strengths and weaknesses, opportunities, and threats to determine the type of strategy. After receiving the data from the water and wastewater system and summarizing them, an evaluation matrix will be prepared. The type of strategy is determined with the help of the internal and external position matrix, with the offensive strategy as the main strategy for the projects. There are three types of sub-strategies introduced for the water and wastewater system with the help of the SWOT matrix.

The researchers are requested to give opinions to rate the attractiveness of the introduced sub-strategies according to the strengths and weaknesses as well as opportunities and threats with the help of the planning matrix. In the matrix prepared for answering, first of all, the method of determining the attractiveness score for each of the strategies introduces by providing a Quantitative Strategic Planning Matrix (QSPM). Then, an example of the matrix is completed in one similar research (Table 1.1).

1.1.3 QUANTITATIVE STRATEGIC PLANNING MATRIX (QSPM)

The matrix is an analytical method for determining the relative attractiveness of strategies. By using this method, it is possible to objectively determine the best strategies.

To prepare the desired matrix, 6 steps must be completed:

- Write down the major external opportunities and threats, internal strengths and weaknesses in the right column of the strategic planning matrix.
- Transfer the coefficients assigned to these factors in the internal and external evaluation matrices exactly to this table in the coefficients column.
- Specify the selected strategies (strategies are pre-specified to facilitate and speed up the response).
- Give the selected strategies an attractiveness score from 1 to 4 (1 = not attractive, 2 = somewhat attractive, 3 = reasonably attractive, and 4 = very attractive).
- Calculate the sum of the attractiveness scores obtained from the product of the coefficients in the attractiveness score.
- Calculate the total attractiveness scores or use the average method. The highest score indicates the attractiveness of the strategies.

Now, in the current QSPM matrix for the pilot, 3 types of sub-strategies have been introduced for the offensive strategy, and the definition of each one and its advantages are mentioned, respectively.

TABLE 1.1 Fuzzy Logic Matrix and Quantitative Strategic Planning

Strategies						Description/ Coefficient	Row	Factors
Advanced Innovative Strategy (1) **Product Development Strategy (2)** **Horizontal Integration Strategy (3)**								
(1)	**(1)**	**(2)**	**(2)**	**(3)**	**(3)**			
Sum of attractive score	**Attractive score**	**Sum of attractive score**	**Attractive score**	**Total attractive score**	**Attractive score**			
–	–	–	–	–	–	–	O1	Opportunities facing the organization
–	–	–	–	–	–	–	T1	Threats facing the organization
–	–	–	–	–	–	–	S1	Strengths of organization
–	–	–	–	–	–	–	W1	Weaknesses of the organization

Total attractive score

O1: Opportunities

T1: Threats

S1: Strengths

W1: Weaknesses

1.1.4 HORIZONTAL INTEGRATION STRATEGY

The purpose of horizontal integration is to implement one of the strategies that the projects increase their control over them. At present, in strategic management, one of the most well-known ways is that projects try to use growth strategies with horizontal integration. One of the common methods is that in many industries, organizations, and projects merge. The features that have been introduced for the merger of projects so far are:

- Optimal use of the available capacity;
- Optimum use of the available workforce;
- Using a phenomenon called "economies of scale."

1.1.5 PRODUCTION DEVELOPMENT STRATEGY OF WATER

The purpose of water product development is the implementation of one of the strategies in which the project tries to increase the efficiency of the system by improving or modifying the current water production and services. This strategy is suitable for projects that are competing in an industry that has very high growth in terms of technology. For example, this strategy can be useful in the following areas:

- Continuous improvement of the quality of water and sewage services to consumers;
- Development of water distribution network and sewage disposal;
- Increasing the amount of water production and sales.

1.1.6 LEADING INNOVATION STRATEGY

Leading innovation strategy has positive features from different points of view for production, which are:

- smartening of water distribution network;
- smartening of water transmission; and
- smartening of water treatment facilities.

This strategy refers to maintaining a leading and innovative technological position in the range of products, being a pioneer, and taking risks in offering new products or entering new projects. For this reason, researchers believe that projects, regardless of whether they are high or low technology, should abandon their traditional (defensive) and imitative roles and turn to more aggressive strategies such as the strategy of leading innovation, which requires establishing water research and development units, expanding research and study activities in the water project to achieve the goals of sustainable development.

Some researchers also concluded that having a strong competitive advantage and developing the competitive position of water companies depends on adopting a leading innovation strategy. Because the leading innovation strategy leads to the improvement of the innovative capability and thereby achieving innovation success or performance improvement. This strategy can be efficient and effective in different fields for the competition of projects.

Projects that use this type of strategy are continuously looking for environmental opportunities and are constantly seeking to test their potential to respond to the emerging trends of the environment and make changes before their competitors. For example, responding quickly to opportunities such as customers' increasing need for safe drinking water and sanitary disposal of sewage, as well as the development of urbanization and housing construction, can be effective through the adoption of this strategy by project managers.

Being a leader in accepting and applying organizational and technological innovations is one of the other features of the leading innovation strategy. For example, the creation and application of new and superior technologies in all areas of software and hardware in the company's facilities can have a positive effect on the company's performance and efficiency. Also, the use of advanced technology with environmental considerations can make it possible to achieve the goals of sustainable development [15–20].

In this work, GIS Ready was provided based on the following procedure in complience with networked sensors, RS, and IoT:

- CAD space to GIS space.
- Fixing errors in the CAD space.
- Convert graphic information from DWG format to SHP.
- Completing the layers of descriptive and spatial information.
- Fixing the errors in the GIS space.
- Snapped together with proper tolerance.
- Eliminating toll errors that are in the wrong place.
- Spaghetti space to topology space.
- Preparing a conceptual model for network modeling in GIS space.

1.1.7 SMARTENING OF WATER DISTRIBUTION, TRANSMISSION, AND TREATMENT FACILITIES

1.1.7.1 WATER TRANSMISSION LINES

It is the pipelines that collect water from wells, surface sources such as rivers or dams and transfer it to treatment plants or ground reservoirs, or cities. These lines are generally outside the urban area, and with the expansion of the city, a part of it may also be placed inside the urban area.

1.1.7.2 INTRA-URBAN WATER TRANSMISSION LINES

It is the pipelines located in the urban area that transfer the required water from the underground reservoirs of the urban service to the distribution network or air sources within the city.

1.1.7.3 DRINKING WATER DISTRIBUTION NETWORK

These are urban water distribution pipelines within the residential areas of cities or water distribution lines that are built in the preparation projects of settlements before their construction.

1.1.7.4 RAW WATER DISTRIBUTION NETWORK AND GREEN SPACE IRRIGATION

These are urban water distribution pipelines within residential areas that are built for the purpose of urban green space irrigation.

1.1.7.5 WATER PUMPING STATIONS

1. **Normal Operation:** In operating the pumping station, the following items must be controlled.
2. **Storage Tanks:**
    - Controlling the water level inside the tanks;
    - Checking the condition of the inlet and outlet valves.

3. **Electric Motors**

- Engine sound control.
- Shell temperature control.
- Control of starting ampere, working ampere, voltage, and engine speed.
- Electronic consumption registration.
- The output cables from the main switches and the start switches of the electropumps should not be touched by hand if the temperature is too high (30C). Turn off the key and inform the repair group.
- Clean the inside of the electric motor with a dry air compressor at a pressure of about 4 atmospheres.

4. **Pumps:**

- Coupling control;
- Controlling the sound of the pump and its vibration;
- Pump shell temperature control;
- Control the circulation of the cooling system, if any;
- Ventilation system control;
- Control of input and output current intensity.

5. **Suction Pipeline:**

- Negative line pressure control;
- Checking the status of the valves (on and off);
- Checking the sealing of valves and fittings.

6. **Transmission Pipeline:**

- Positive line pressure control;
- Checking the status of faucets and connections;
- Checking the sealing of faucets and fittings.

7. **Security and Lighting System:**

- Controlling the status of security systems in the facilities;
- Controlling the status of the whole system by setting up daily reports;
- Ensuring the lighting of the pump house during the day, controlling it, and taking action to fix any defects to prevent issues at night.

1.1.7.6 FAULT FINDING AND FIXING IT

If periodic services and repairs are carried out according to the schedule, the workload of the repair unit will be greatly reduced. In any case, there are reasons to form an experienced construction group at the pumping station, to fix the defects and prevent disruptions in operation. Defects and faults created in the electric motor of the pumps can have hydraulic or mechanical reasons.

1.1.7.7 PROBLEMS

Air has entered the pump:

- The pump or suction line is not completely filled with water.
- The suction height is high.
- The suction pressure is equal to the water vapor pressure.
- The presence of air bubbles in the suction pipeline.
- The inlet of the suction pipeline is not completely in the water.
- The pump speed is very low.
- The pump is rotating in the wrong direction.
- The system overhead exceeds the design overhead.
- Operating the pumps in parallel is not correct in the operating conditions
- The presence of foreign materials in the pump impeller

1.1.7.8 PUMP WITH LESS CAPACITY

The pump or suction pipeline isn't completely filled with water.

- The height of suction is more than allowed.
- There is water, air or other disturbing gases inside.
- Air infiltration into the pump suction pipe.
- Air infiltration into the pump chamber.
- The suction pressure is equal to the water vapor pressure.

The one-way valve at the end of the suction pipeline (suction valve) has been pinched or seized.

- The rotation speed of the pump is low.
- The system overhead exceeds the design overhead.

- The viscosity of the fluid is higher than the design value.
- Parallel operation of pumps is not suitable in operating conditions.
- The pump impeller has foreign materials in it.
- Damage to the pump impeller due to corrosion.
- Deterioration of the sealing gasket

1.1.7.9 CAUSES OF DEFECTS IN CENTRIFUGAL PUMPS

1. **Problems and the Reasons:**
   - The pump's rotation speed is low.
   - The direction of rotation of the pump is wrong.
   - The system overhead exceeds the design overhead.
   - Fluid viscosity is different from the design value.
   - Corrosion and damage to the pump impeller
   - Deterioration of pump sealing washers
   - The pump head is insufficient
   - The presence of air bubbles in the incoming water
2. **Packing Soon:**
   - The water supply pipe (water seal) is closed.
   - It breaks down
   - The pump is installed flush.
   - Bending of the pump axis or shaft
   - Shaft corrosion
   - Installation of wrong packing
   - The type and material of packing is not suitable for working conditions
   - Engine misalignment
   - The flower is tightly closed.
   - The distance between the shaft and the body
   - The presence of foreign substances in the sealing liquid
3. **The Pump Overheats:**

   i. **The Pump is Not Aerated:**
      - The difference between suction and steam pressure is small.
      - The pump is operated with less capacity.

- The parallel operation of the pump is not suitable in the operating conditions.
- Uneven pump installation.
- Abrasion of moving parts on fixed parts of the pump.
- Axis or shaft bending.
- Imbalance of the moving engine.
- The speed of pump has reached critical speed.

ii. The Pump Vibrates or Makes Noise:

- Suction pressure is the degree of water vapor pressure.
- The suction line valve is too small or blocked.
- The inlet of the suction pipe is insufficiently located in the water.
- It is operated with less capacity.
- The pump or suction pipeline is not completely filled with water.
- The suction height is high.

4. Problems and Causes of Defects in Centrifugal Pumps:

i. Problems and the Reasons:

- Uneven installation of pump and drive motor.
- The foundation is not strong enough.
- The pump shaft is bent.
- The pillow is a moving part to a fixed part.
- Damage to the pump impeller.
- Lack of lubricant.
- There is rotation in the pump axis.
- The pump impeller has foreign materials in it.

ii. Power Consumption is High:

- The drive motor works in reverse direction.
- The head of the whole system is more or less than the design head.
- The difference between the specific gravity and the viscosity of the fluid with the design value.
- Foreign materials found in the pump impeller.
- Uneven installation of pump and motor.
- Bending and deviation of the pump shaft.

- Rubbing a moving part on a fixed part.
- Wrong packing installation.
- The type and material of packing are not suitable for working conditions.
- The flower is tightly closed.
- The speed of the era is very high.

5. **Overturning the Pump:**

   i. **The Pump is Not Ventilated (Cavitation):**

   - There is water, air, or other gases inside.
   - The presence of air in the suction line
   - The inlet of the suction pipe is not completely in the water.
   - The water pipe is flooded.
   - The suction height is high.

6. **Bearings Break Down Quickly:**

   i. **Misalignment of Pump and Motor:**

   - Rubbing a moving part against a fixed part;
   - Dry working of bearings;
   - Incorrect installation;
   - The presence of foreign materials in the bearing;
   - Bending of the pump shaft.

   In the first case, the pump cannot pump, or the pump operates with less than its nominal capacity or does not provide enough pressure. In the second case, the amount of electricity consumed increases, and problems such as the valve of the pump causing noise, corrosion, and breakage of parts. In general, any defects and problems in the operation of pumps can be caused by one or more reasons. By finding the cause, the repair department manager can easily determine the percentage of its resolution [21–29].

7. **The Problems and Causes that Generally Occur in Centrifugal Pumps:**

   i. **Maintenance and Lubrication:** Maintenance principles of pumping stations. The maintenance of a pumping station, like any industrial system, is divided into three parts:

   - Continuous control of system details;

- Periodic services; and
- Period replacements.

In a pumping station, the details of the station must be continuously controlled by the maintenance group. They need to find system faults, predict the causes, and fix them. The periodical services of a pumping station for each of the constituent parts usually include the following:

- Control of gate valves and one-way valves.
- Lubrication of bearings of electric motors and pumps.
- Opening and closing the pumps and checking the condition of the shaft, impeller, and internal wall of the pump shell.
- For every 2000 working hours of the pump, the bearing grease should be replaced.
- Starting amp test, working amp, head of electric motors.
- Service of electrical panels and sensitivity testing of safety and protective circuits.
- Service of control systems such as limit switches and flowmeters.
- Service and control of loose couplings.

Periodic replacements of a pumping station for each of the components in a way that can include the following:

- Bearings and roller bearings (pumps and electric motors);
- Rings and graphite (every 6 months);
- Pump impeller;
- Pump shaft;
- Ball shaft of valves;
- Manometers;
- Couplings;
- Water flow measurement meters (if available);
- Sensitive parts of electrical panels.

1.1.8 PERIODIC OPERATION OF INSPECTION AND MAINTENANCE OF THE PUMPING STATION

1. **Daily:** All defects observed during hourly and daily visits in the operation of the pumps, such as changes in the sound of the pump

motor, sudden changes in the temperature of the pump shell, leakage from the pump, pressure, and flow readings, are observed by the operators.

2. **Monthly:** In addition to the daily visits by the operators, the following visits and operations must be performed by the construction team on a monthly basis:
    - The state of packing and sealing gaskets should be checked and replaced if necessary.
    - Adjusting and aligning the pump motor, or in other words, inspecting and measuring the distance of the couplings.
    - Greasing and lubrication of electric motor bearings with 160? fireproof grease.
    - Inspection and repair of pipes and connections if needed.
    - Inspection and repair or replacement of pressure gauges if necessary.
    - Pump electromotor vibration control.
    - Cleaning the inside of the electric motor of the air compressor belt pump with a pressure of about 4 atmospheres.
    - Investigating the condition of the bolt and gland
3. **Annually:** The measures that should be taken during the annual visits are as follows:
    - The pump components should be opened, washed, and cleaned.
    - Descaling of impeller and pump shaft.
    - Lubrication of bearings and roller bearings.
    - Painting faucets, fittings, and pump shell.
    - Inspect the floor and walls of the pump house and repair them if any defects are found.
4. **Operation of the Unit:**
    - Close the valve located on the thrust pipeline of the desired pump.
    - Turn off the engine.
    - Close the valve located on the suction pipeline.
5. **Safety:**
    - Electropump should never be used outside the specified range in terms of water flow.
    - Motor speed.

- Pressure and temperature.
- Avoid keeping the electropop on for a long time while the outlet valve is closed.
- Under no circumstances should the pump work without load.

1.1.8.1 FAUCETS OF THE DISTRIBUTION NETWORK

The valves in the urban water distribution network are considered key control points of the network. The absence of a suitable plan and method for the maintenance and operation of the faucets of the distribution network often causes them to lose their desired and necessary performance after a period of installation. Instead of being used as executive management tools, it causes problems. The first step to repairing and maintaining these units is to know their function and role in the distribution network. After that, the existing obstacles should be removed and their protection and maintenance should be known, and then with the schedule of periodic visits and service, while applying the necessary controls, repair, and maintenance should be considered.

1.1.8.2 GATE VALVE

This valve is used to completely shut off and connect the water flow in the distribution network. These valves are manufactured and marketed in different sizes and to withstand different pressures. The material of these valves is generally cast iron or ductile iron and they are widely used in distribution networks and water transmission lines. A gate valve, along with its components, has the most common installation points:

- The intersection of pipelines.
- At the entrance and exit of tanks.
- In water transmission lines known as taps.

There are two types of the valve in terms of how it works:

- A valve with a rising stem (RISING STEM VALVE) has a stem or rod that also rises, allowing water to pass.
- Gate Valve with a fixed stem (NON-RISING STEM VALVE) moves on a rod or stem in this group. These valves are generally used for underground installation.

- **Advantages of Gate Valve:** When it is completely open, the water pressure drop is very small and can be ignored.
  - Long life span;
  - High work pressure;
  - There is less water leakage in them.
- **Disadvantages of Gate Valve**
  - In large diameters and for high pressures, it is difficult to open and close it by hand, and the valve must have a passage.
  - If the gate valve is not operated (opened and closed) regularly, due to the accumulation and deposition of particles in the water, opening and closing them will not be easily and correctly done.

1.1.8.3 REPAIRS

Valve valves often need to be repaired due to the following reasons:

- Deterioration of the sealing thread (PACKING).
- Corrosion and rusting of the tap rod or stem.
- Corrosion of tongue or disc.
- Deterioration of sealing rings.

Each of the above causes malfunctions and leakage from the valves. In order to fix the defect and repair each of the above-mentioned items, it is done in the following way:

i. **Replacing the Sealing Thread:**
   - Open the gland.
   - The sealing thread must be removed from the sealing chamber (STUFFING BOX) and its place should be cleaned.
   - The thread of the new seal should be placed in place and secured in place with a gland.
   - Pour a few drops of oil on the sealing thread and then put the valve gland in its place and close it tightly.

ii. **The Service of the Bar or Stem of Valves:**
   - Open the gland.
   - Take out the sealing thread.
   - Open the cap and remove the rod.

- Clean the surface of the rod with sandpaper or sanding oil, and if there is a lot of corrosion and rust, the rod should be replaced.
- Put the rod inside the cap.
- Seal gaskets should be replaced.
- The cap is firmly in place.
- Replace the sealing thread.
- Open and close the valve.

iii. Fixing the Corrosion of the Disk of the Valve and Sealing Ring:

- Open the gland and cap.
- Remove the rod.
- Remove the disk and clean it with sandpaper, and replace the disk if there is a lot of corrosion.
- Clean the sealing rings and replace them if necessary.
- Reassemble the valve parts. Replace sealing gaskets and sealing thread.

The gate valve should not be left open or closed for a long time, because the water inside the valve is filled with salts when it is open, and the valve is not completely closed and sealed. In the case of being closed for a long time, the solutes in the water accumulate around the valve drawer and cause a lot of pressure on the valve axis when opening the valve, in which case there is a possibility of the axis being cut. It is necessary to maneuver the valve at least once a month.

1.1.8.4 CHECK VALVE

A one-way valve is a valve that is installed in the pipeline. When water flows, the moving force of the water opens it, and when the water flow changes its direction, the plate of the valve closes the pipe. In general, one-way valves work automatically and prevent reverse flow in pipelines. One-way valves are generally divided into two main groups: hinged one-way valves (LIFE CHECK) and one-way ball checks (BALL CHECK). Hinged one-way valves are often used when the goal is to pass the flow completely, and it is the most common and most used one-way valve. The one-way valve is used in cases where the valve must operate frequently and in networks with high pressure and high flow rate. Among the factors that cause defects and disruptions in the operation of the one-way valve are the corrosion and wear of the thorn holder, cap maintenance, valve deposition, and disc corrosion. In order to fix such defects, proceed as follows:

- Open the valve cover.
- The nut or spine of the axis should be removed.
- The valve disc should be sanded or ground.
- In case of damage to the axis, replace the axle pin and nut and their cover.
- Mount the valve components again.
- Valve should be tested.

1.1.8.5 BUTTERFLY VALVE

The construction of butterfly valves is simpler than other valves. These valves have a part in the shape of a rotating plate or disk that moves around an axis or a rod or a stem. Opening and closing these valves is simple and is done only by turning the disk so that it fills the entire width of the pipe. Depending on its type, the plastic lining of the seat of these valves is made of adhesive material, or it is fixed to the disc with mechanical devices, or it is embedded in the disc, or it is made of a mechanical spring seal or a metal seat that is placed in the form of an O-RING around the edge of the disc. Among the advantages of these valves are:

- The coefficient of water passage in it is directly related to the opening percentage of the valve, and therefore it can be used to control the flow and create a pressure drop in the network.
- Opening and closing it is done with more ease and speed.
- It needs less space for installation.
- Its repair and maintenance cost is relatively low.

These valves are easy to maintain and generally do not need lubrication until the stem O-ring is replaced. The disc and stem, as well as the O-ring disc and its plastic seats, are easily replaceable. Replacement of the plastic metal lining of the sealing surfaces with adhesive materials cannot be done on-site and is done only by the manufacturer.

1.1.8.6 AIR VALVE

An air valve is a device to let the air out of a pipeline or to let air into it. Air valves are divided into three categories:

- Single-hole air valves, with a small hole: These are used to release air from the pipeline that is under hydraulic work pressure.

- Single-hole air valves, with a large hole: This valve is used to ventilate and empty the pipe. When the pipeline is filled with water, the air is removed from it, and when the pipe is emptied of water, air enters the pipeline. When the pipeline is under hydraulic pressure, the above valve is not able to release the small amount of collected air, and in this sense, the hydraulic pressure prevents the ball from separating from the hole.
- Two-hole air valve: This performs the work of both the above valves. In terms of operation, if the valves are installed in the right place, they will not have any particular problem. Of course, it is necessary to visit it periodically, and if its opening is clogged with sediment, it should be descaled and rusted.

1.1.9 PERIODIC VISITS, REPAIR, AND REPLACEMENT OF VALVES

The fact is that during the use of the distribution network, there is no justification for cutting off the water for a long time. Therefore, it should not be enough to fix problems as corrective maintenance. Rather, it is necessary to optimize the use of the facilities by training and equipping the crews and conducting periodic visits and services.

Each of the valves used in the network, even if they are made of the best metal materials, will suffer over time and by corrosion, rusting, and sedimentation become clogged. Like other mechanical systems, failure to maneuver the valve on time will gradually make it impossible to maneuver the valve. So that their maneuver is recommended twice a year. On the other hand, the access valve to the line valve due to the spread and various implementation operations in the city such as drilling, surface changes, and asphalt may be invisible, and as a result, the possibility of valve maneuvering is lost. The more important point is that each faucet is installed for a reason and to connect and disconnect the water in a specific area. If a pipe breaks due to drilling, leakage, or as a result of any kind of accident, by closing the valve of the relevant line, the outflow of water from the accident site is controlled. Therefore, if one of the valves in the network becomes inoperable or unmanageable, at least several other line valves should be closed instead, which will cause the water to be cut off in a wider area, and water will be wasted for a long time. In addition to the mentioned points, if any of the valves in the line are not opened and closed continuously, and if it becomes corroded, it will automatically lose its water content after checking and will leak from the outside, resulting in water wastage and infiltration. As a result, the soil settles under the pipe and breaks the pipelines. If there is no sealing valve, the repair of the accident pipe is impossible in some cases. A pump is needed to drain the water from the accident site. In the maps of the water distribution network, which have been updated with a local survey, all the visible and invisible taps in the network have been identified and reflected on the maps. According to the fact that pipes are considered to be part of the main pipes of the distribution network, taking into account the common standards, it has been recommended that the faucets related to these lines have basins. To be able to inspect periodically and to ensure the correct operation of the main facets of the network, when there is a break in these pipes, and to reduce water losses caused by this, they should be placed inside the basin.

1.1.9.1 FAUCETS

Faucets that are generally used in urban water distribution networks include gate valves, butterfly valves, one-way valves, and air valves. When determining the standard dimensions of the faucet basin, it is necessary to consider the technical and standard dimensions of these faucets. The technical dimensions of these valves are presented when introducing them. It is important to note that the technical dimensions may vary as different manufacturers produce and market their products with slight differences.

1.1.9.2 GATE VALVE

This valve is used to shut off and connect the water flow in the distribution network. Gate valves are manufactured and marketed in different sizes to withstand various working pressures. The material used for these valves is typically cast iron or ductile iron.

1.1.9.3 BUTTERFLY VALVE

The structure of this valve is simpler than other valves, especially gate valves, and they are available in two types: flanged and wafer. Butterfly valves are used to control and regulate the amount of flow passing through a pipeline.

1.1.9.4 ONE-WAY CHECK VALVE

This valve is used to prevent backflow in a pipeline. These valves have a simple structure and their material is cast iron or ductile iron. These valves are made in various forms such as hinged, valved, and ball valves.

1.1.9.5 AIR VALVE

Air valves are among the valves that facilitate the entry and exit of air inside a pipeline. These valves are manufactured and marketed in different shapes and types such as single orifice and double orifice. Unlike other valves, the air valves are not located in the pipeline route but are mounted on it. Therefore, air distance and side distances are important for them in the construction of ponds.

1.1.9.6 WATER DISTRIBUTION NETWORK PIPES

The skeleton of the networks is made up of pipes. These pipes can be used in sizes of 60, 80, 100, and 150 mm for subsidiary networks and up to 1,250 mm in diameter for main networks.

Pipes are the only part of the urban water network that are always buried underground, so they are more vulnerable than other parts of the network and will have accidents. Therefore, maintenance of the network in the sense of visiting them is not possible only when accidents occur, but it is better to carry out leak detection by a leak detection device during a regular program and according to the priority of the areas according to the age of the network and its importance. The network should be repaired before causing any massive and costly damage. In this regard, it is necessary to mention that usually the amount of water wasted from small leaks in the network is more than from large leaks that cause accidents [30–39].

The reasons for reducing the water transfer capacity of pipes are:

- Foreign objects inside the pipes, such as wood, stone, etc., that enter the pipes during pipe laying or necessary repairs.
- Sedimentation of sludge, iron, manganese
- Creating a coating layer of sedimentary materials on the inner wall of pipes and protrusions caused by rusting or water corrosion or lime deposits.

- Air accumulation in the highest points of the level
- Reducing the cross-sectional area of the pipe due to sedimentation of faucets in semi-closed states or bending of pipes

Therefore, to prevent large losses and damages, the accurate implementation of the network and the use of quality materials, and the careful monitoring of its implementation while observing the technical principles of implementation are factors that can have a great effect on the health of the network and increase its life.

Otherwise, in conditions where it is possible that the water transmission through the pipes is reduced, the following measures should be taken:

- Full air evacuation control throughout the length of the pipeline.
- Washing pipes.
- Visiting the accessible parts of the pipe to control the sediment layer created on the surface of the inner wall of the pipes.

1.1.10 TANKS

1.1.10.1 PRE-LAUNCH

Before starting the exploitation operation, it must be ensured that the following things are fully fulfilled:

- All the problems that caused the failure of the unit are recorded in the operational sheets.
- The repair and replacement of the parts that were involved in the occurrence of problems have been done.
- The tank overflow system is ready for operation.
- The performance of the faucets entering the tanks is guaranteed.
- The performance of the outlet taps from the tanks is guaranteed.
- The performance of the chemical injection and chlorination system is guaranteed.
- Preparing the daily report book (operational sheets)

1.1.10.2 NORMAL OPERATION

- Continuous control of the height of the water level in the tank and recording it in the relevant worksheet.
- Ensuring that the inlet and outlet valves are open.
- Ensuring the performance of ventilators and ventilation system.

- Inspection and notification of the absence of leaks in the walls and roof.
- Ensuring the operation of inspection valves
- Visiting the appearance of water
- Record the report of each of the above cases in information sheets
- Continuous control of chemical injection rate

1.1.10.3 CONNECTIONS

Connections that are generally used in place of the faucet basin for the operation of various faucets.

1.1.10.4 VALVE MANHOLES

The standard technical dimensions of the faucet basin are calculated and presented based on the faucets and equipment used in the urban water distribution network in different situations. Regarding the dimensional details provided, the following points can be mentioned:

- The height of the ponds is considered variable due to the difference in the depth of piping.
- The minimum standard dimensions of the number of ponds.

1.1.11 SMARTENING OF WASTEWATER COLLECTION, TRANSMISSION, AND TREATMENT FACILITIES

The wastewater produced for different periods and the status of water supply and wastewater disposal facilities in the existing conditions, preparing a map of the status of the existing traditional network of the city, examining the types of wastewater collection networks, determining the most suitable network according to the conditions of the city, comparing the technical and economic types of water pipes and sewage, determining the optimal type and diameter, and finally checking water and wastewater treatment methods, process calculations of selected treatment methods, determining the overall dimensions of treatment plant units, and how to implement executive modularization during the design period, and finally cost estimation is done.

1.1.11.1 WASTEWATER URBAN PIPELINES

1. **Wastewater Collection Network:** The wastewater collection network consists of urban wastewater collection pipelines, pre-construction settlements, and their connection to treatment plant systems and disposal sites.
2. **Pipeline:** The pipeline generally consists of the following items:
   - Pipe;
   - Pipeline connections;
   - Accessories;
   - Valves;
   - Accessories;
   - Basins and manholes.

There are other related structures that are used in the complete construction of the pipeline. The construction of the pipeline as a hydraulic structure includes the open trench and the foundation. It also includes the bed, pipe installation, first stage embankment, hydraulic testing, final embankment, washing, and disinfection. The pipeline installation is finally completed by finishing works such as the construction of ponds, manholes, and the installation of valves for the complete implementation of the work.

1.1.12 TYPES OF SEWER PIPES

1.1.12.1 FIBERGLASS PIPES

Fiberglass pipes were first produced in 1948. The first use of these pipes, which is still the main consumer, is the oil industry. According to the type of resin used and how they are reinforced, fiberglass pipes are divided into several types, one of which is synonymous with the other and is known as G.R.P. In this technical specification and for convenience, the names fiberglass and G.R.P. refer to pipes with a composite structure of resin, glass fibers, and other materials. These pipes have the same concept in G.R.P. Pipes are used in applications such as water transfer, sewage collection, surface water collection, seawater transfer, irrigation, and drainage networks, industrial wastewater, power plants, oil and petrochemical industries, chemical industries, etc.

1.1.12.2 POLYETHYLENE PIPES

Polyethylene is prepared from the polymerization of ethylene and olefin, along with antioxidant carbon black. This material is commercially available in three types: low specific weight, medium specific weight, and high specific weight. The higher the specific weight of polyethylene material, the greater its resistance. The design of pipelines should be done with attention to different conditions at any point in time. In the production of polyethylene pipes, an extruder machine is used. The machine automatically heats these materials into a paste and passes them through a special mold with pressure. After cooling, the material comes out of the machine in the form of a pipe. Polyethylene pipes are resistant to the effects of chemicals, so they do not need an external coating against the corrosive effects of the soil in which the pipe is buried, or an internal lining to resist the corrosiveness of the fluid passing through the pipe.

1.1.12.3 PVC PIPES

The group of materials named hard PVC is one of the most important materials made in the last few decades. By adding softener to PVC, the soft pipe is produced. The properties of PVC pipes come from their raw materials. PVC resin, which is the main component of PVC pipes, is a polymer made from gas or oil along with salt water and air. After manufacturing, PVC resin is mixed with thermal stabilizers, oily materials, and other compounds and converted into PVC pipes with extruder machines. The properties of PVC pipes are desired from several aspects, which include the type of base resin used, impact resistance, tensile strength, modulus of elasticity in tension, and deformation due to heat. In addition to that, the degree of resistance against chemicals and acids is also a factor in the selection of PVC pipes.

1.1.12.4 ASBESTOS CEMENT PIPES

They are produced from a mixture of Portland cement with asbestos fibers (fireproof cotton). This kind of pipe has been used in America since 1931 and several years before that in other countries, especially Italy. Due to the suitable characteristics of asbestos cement pipes, their use in various fields was quickly welcomed. Asbestos cement pipes are widely used in both transmission and distribution lines under pressure and gravity lines. The wall thickness of pressed cement asbestos pipes is a function of internal pressure and pipe diameter. Therefore, the greater the internal pressure or diameter of the pipe, the thickness of the pipe wall increases proportionally.

1.1.12.5 REINFORCED CONCRETE PIPES UNDER PRESSURE

Used in water supply projects, these pipes are usually employed in water transmission lines. They are less commonly used in water distribution networks, and their use in such networks is typically limited to water supply lines of distribution networks with a diameter of more than 300 mm. These pipes function as a kind of transmission line within city boundaries, branching off and not directly connected to them, but responsible for distributing water from service tanks to the pressurized areas of the city.

1.1.12.6 ORDINARY CAST IRON PIPES UNDER PRESSURE

Ordinary cast iron pipes, made of gray iron, have been one of the oldest types of pipes used in transmission lines and water distribution networks in various countries. The main advantage of this type of pipe is its high resistance to soil and water corrosion, resulting in a significant lifespan. However, due to the fragility and heavy weight of these pipes, especially for large diameters, transportation problems and high costs for transportation and installation can arise.

1.1.12.7 DUCTILE CAST IRON PIPES

Ductile cast iron pipes, due to their remarkable advantages in terms of flexibility and resistance to impact, pressure, corrosion of water and soil, as well as less weight, have quickly replaced ordinary cast iron pipes in the last few decades. Therefore, in this specification, after mentioning the standards, only ductile cast iron pipes are discussed.

According to the manufacturing technology, steel pipes include three general types as follows:

1. **Seamless Steel Pipes:** Seamless steel pipes are produced in small diameters through hot rolling and are used for high-pressure work. Therefore, such pipes are less used in water transmission lines and water distribution networks, and urban facilities.

2. **Straight Seam Pipes:** Straight seam steel pipes are produced by changing the shape and piping of flat steel sheets, and they are made in different thicknesses. The shape of the pipe is changed by the press, resulting in a seam that is initially O-shaped and then U-shaped in this method. First, the flat sheet of steel is welded by the contact of two edges of the sheet along the length of the pipe.
3. **Spiral Tubes:** Spiral steel pipes are made by twisting and rolling steel strips using a special machine and inside the relevant mold. In this method, the seam is formed between the two edges of the two sides of the sheet that spirals around.
   - **Piping Trench:** It is a channel that is built according to the routes determined based on the executive plans and these technical specifications, in order to install pipes, connections, and accessories.
   - **Altitude Elevation Code:** It is the height of the desired points compared to the average sea level according to the reference coordinates (B.M) or the height of the points resulting from the harvest according to the elevation code of local reference points (NCC) mapping.
   - **The Base Level of the Pipe:** It refers to the elevation code of the lowest point inside the pipe, not at the joint.
   - **The Top Level of the Pipe:** It refers to the elevation code of the highest point outside the pipe, the crown of the pipe, not at the joint.
   - **Natural Ground Level Amount:** It refers to the land elevation of the pipeline route and related facilities before the start of construction work.
   - **Finished Level Amount:** It refers to the elevation of the land for the piping route and related facilities after the completion of the work.
   - **Trench Floor Level:** It refers to the elevation under the soil of the trench floor.
   - **Depth of the Trench:** The depth of the trench is the difference in height between the bottom of the trench and the average elevation of natural or modified land above the trench on both

sides. When there is a need to build a foundation under the pipe bed, this depth increases accordingly.

- **Width of the Trench:** The width of the trench is the width of the bottom of the trench.
- **Longitudinal Slope:** Longitudinal slope means the ratio of the difference in elevation under the pipe between the two desired points to the distance between these two points.
- **Trench Foundation:** The purpose of the foundation is to modify the ground under the trench floor of the piping route using desired and acceptable methods to improve and stabilize it when passing through lands with poor quality soil, materials, mud, and the like.
- **Trench Bedding:** The purpose of trench bedding is to prepare the trench floor with selected and acceptable materials to improve and stabilize it.
- **Rigid Pipe:** It is a pipe that does not change its shape under external load and before breaking, such as asbestos cement pipes, unbreakable cast iron, and concrete pipes.
- **Flexible Pipe:** It is a pipe that changes its shape under external load and before breaking, such as plastic and steel pipes.
- **Diameter of Pipes, Accessories, and Valves:** The diameter of pipes, accessories, and valves is their nominal diameter, which is defined in international standards or manufacturers' tables.
- **Connectors:** These are parts and accessories that connect two pieces of pipe with their accessories, such as sleeves, glands, mechanical couplings, washers, and flanges.
- **Accessories:** The accessories are devices and accessories that either connect the pipeline to the valves or are used to change the piping route, such as elbows, tees, conversions, etc.
- **Pipeline Boundary:** It is the minimum width of the boundary where the pipe can be easily threaded. The installation and movement of pipe-laying machines can be carried out easily, and finally, the implementation of related earth operations including leveling and preparation of the boundary, digging trenches, and

depositing the resulting soil. Excavation, re-filling, and transportation of excess soil can be done easily.

- **The Scope of Road Operation:** The width of the scope of operations is the width of the desired part of the pipeline boundary, which is used for the transportation of machinery, threading the pipe, digging the pipeline trench during the implementation of the operation, and also maintaining the line. In this method, the pipe is in operation.
- **Selected Soil:** It means selected soil or high-quality soil for re-embankment under, around, and on the pipe or pond, which has the desired characteristics and is free of organic matter, frozen soil, large stones, tree branches, and roots.
- **Stony Land:** It is land that requires the use of incendiary and explosive materials to dig or extract it or the use of heavy machinery such as bulldozers.
- **Collapsible Land:** It refers to land that is not easily dug according to the plan due to the fall of the soil around the trench or pit, and to dig it, special measures such as stakes or shielding must be done in it, or the trench with a suitable slope should be dug to prevent the total fall of the soil.
- **Normal Land:** Land outside the definitions of stony land and collapsible land is referred to as normal land. The definition of the desired land in this technical specification is for piping works and the construction of ponds or dams [40–49].

1.1.13 WASTEWATER TREATMENT PLANT

The design of the equipment should be based on the smart method. Its parts are replaceable and facilitate its inspection during operation, cleaning, lubrication, and repairs. It minimizes the risk of fire and prevents the entry of dust or insects and is completely safe in terms of electrical, mechanical, and chemical operation, and minimizes the installation costs of the equipment.

1.1.13.1 ACTIVATED SLUDGE PROCESS – AERATION

The process of the wastewater treatment plant is of the activated sludge-aeration type, and the process units in the treatment plant are as follows:

Process Units:

- Mechanical scarab;
- Manual scarab;
- Aerated grain harvesting units;
- Ultrasonic flow measurement (UFM) device;
- Aeration tank and selector;
- Secondary sedimentation basin;
- Flume parcel (before chlorination);
- Chlorine contact tank;
- Sludge disposal and return pumping station;
- Concentrated sludge pumping station;
- Sludge thickening tank;
- Sludge dewatering unit;
- Bastar water drain pumping station.

1.1.13.2 QUANTITATIVE AND QUALITATIVE FOUNDATIONS OF THE REFINERY

The quantitative and qualitative basics of the sewage treatment plant.
Description:

- The number of purification modules (Module).
- Capacity of purification modules (person).
- Sanitary sewage per capita (LPCD).
- Average sanitary sewage (m^3/d).
- Average seepage and infiltration waters (m^3/d).
- Average sanitary sewage and seepage (m^3/d).
- Minimum current coefficient.
- Maximum current coefficient.
- Maximum amount of sewage entering the treatment plant (m^3/d).
- The minimum amount of wastewater entering the treatment plant (m^3/d).
- Per capita organic load entering the refinery (gr/ppcs).
- Per capita load of suspended solids entering the treatment plant (gr/pcd).
- The concentration of organic materials entering the treatment plant (mg/l).
- The concentration of suspended substances entering the treatment plant (mg/l).

- Estimated population (without utilization factor) (person).

The flow entering the treatment plant will be the basis of hydraulic calculations.

Type of the Units:

- Average sewage flow (L/S);
- Minimum sewage flow (L/S);
- Maximum sewage flow (L/S).

1.1.13.3 THE MAIN UNITS OF THE REFINERY

The units that play a key role in the operation of the wastewater treatment plant are as follows:

- Dirt stuck;
- Grain collecting pond;
- Device for measuring the incoming sewage flow (Parshall Flume);
- Current divider;
- Aeration pond;
- Final sedimentation basin;
- Telescopic milk hole;
- Return sludge pumping station and extra sludge;
- Sludge thickening pond;
- Sludge drying unit (sludge drying bed or sludge pool or strip filter);
- Device for measuring the outflow of sewage;
- Chlorination pond;
- Unit for removing water from sludge;
- General characteristics of the control system;
- Precision instruments of the wastewater treatment plant.

1.1.14 SMART SYSTEM OF THE WASTEWATER TREATMENT PLANT

To make the best use of the wastewater treatment plant equipment, it is necessary to install a collection of garbage collectors, grit collectors, aeration, secondary settling, sludge treatment units, sludge pump houses, and side units with a proper control system and precision instruments. In addition to facilitating and optimizing the operation of different units and the sewage treatment process, this system should provide the necessary protection for the optimal functioning of the devices and equipment installed in different parts of the treatment plant. The control system should be designed in such a way that each unit has local control and can be installed in the local control panel of each collection. Also, a central control panel will supervise all the collections. Each of the local control units should be equipped with proper control equipment and precise tools to control the devices related to it and also to measure the required values or communicate with the central control room. The control system considered for each of the local control units should function in such a way that all the electric valves, motors, and motor pumps located in each of the complexes, in addition to the possibility of starting from the place of installation of the equipment, can be controlled from the panel. The central control is fully monitored and can be started or stopped remotely. In such a way that their complete status is shown on the spot in the central control room on the Mimic Diagram board. The control method in each set will be done automatically or manually. The Mimic Diagram board includes all the signal lamps indicating the status of different parts of each set with normal signal lamps (steady light) and the error condition is indicated by flashing lamps [50–55].

1.1.15 GENERAL SPECIFICATIONS OF CONTROL EQUIPMENT AND PRECISION INSTRUMENTS

In general, it is necessary to predict the standards and technical specifications of the control equipment and precision instruments of the sewage treatment plant in such a way that this equipment has the best ability and working capacity in the climatic and geographical conditions for which it is designed.

- The specifications of the construction and use of the equipment should be based on one of the valid international standards.
- Valid international standards should be used in the design of PI maps.
- Supply voltage should be selected according to the needs of the design and control system.
- The output signals of transmitters of precision instrument equipment.
- All sensors and transmitter devices are expected to be of mechanical, electromechanical, or electronic types according to the needs and general conditions of the project.
- All indicators, controllers, and constants should be designed to minimize errors and maximize durability in the harshest working conditions.
- All precision devices should be installed on the central control panel as required by the design, in addition to being installed on site.

- All indicators and constants should be selected in standard dimensions, with non-standard dimensions being acceptable if necessary.
- All the control equipment and precision instruments located in the open space must be equipped with thermal or cooling covers so that they do not become damaged due to cold, humidity, etc.
- All primary elements and transmitters related to the measurement of pressure, flow, and height of the liquid surface or other precise elements that are somehow in contact with sewage or liquid must be protected and selected in such a way that they have sufficient durability in terms of rust and corrosion.
- Connections and equipment of precise and measuring instruments on process pipelines should comply with valid international standards.
- The degree of accuracy of all the precision and measuring devices used must meet valid standards.
- The quality of the communication cables of precision instruments and control equipment should be ensured in such a way that the transmission of the required signals is carried out with the greatest possible accuracy over the determined distances.
- Suitable indicators and constants, if necessary, should be considered for local control boards.
- Each of the precision measuring devices must have a suitable plate with at least the following information recorded on it.
- The characteristic sticker of the device.
- Manufacturer's name.
- Model or type and a serial number of the device.
- Measuring range and its accuracy.
- Materials that make up the parts in contact with the liquid.
- Supply voltage and frequency.
- Output voltage or current.
- The conditions of use of the precise device depend on the geographical conditions of the place.
- Type of stimulus for faucets.
- Calibration label.

1.1.16 GARBAGE COLLECTION UNIT

To separate the coarse and floating objects in the sewage, with the aim of preventing malfunctions of overflows, valves, and aerators.

1.1.16.1 THE MECHANICAL, AUTOMATIC GARBAGE COLLECTOR

A manual garbage collector is installed in the adjacent channel so that in case of problems in the work of the mechanical garbage collector, it enters the circuit and can be replaced by the mechanical garbage collector until the problem is solved. Mechanical garbage collector units and manual garbage collector units will be used. In the first stage of operation, only one mechanical and manual garbage collection unit will be built. The technical specifications of the channel before garbage collection for the maximum flow rate at the end of the project are as follows:

- The width of the channel, meters;
- The flow rate, meters per second;
- The height of the fluid (m);
- The slope of the channel;
- Flow speed meters per second.

Description of the units:

- Medium duty (Lit/s);
- Maximum speed in the scavenger (m/s);
- The distance between the bars (mm);
- The thickness of the bars (mm);
- Installation angle (Degree);
- Garbage channel width (m);
- Input channel width (m).

The main equipment of the garbage collection unit:

- Scraber networks;
- Garbage frame;
- Garbage collection basket;
- Garbage collector;
- Complete drive and control system;
- The complete system for transferring and emptying the collected garbage to a silo or silicone for garbage collection uses a conveyor system.
- The water drain will be emptied from the inside of the silo or trash storage solution.

Special technical specifications for entry garbage collection equipment.

- Garbage bars are made of steel and installed in the channel as straight bars without arcs at a suitable angle.

- Corrosion protection is achieved through hot galvanizing, sandblasting, zinc-rich lining, and epoxy paint.
- The garbage net should have a sufficient number of reinforcements and necessary connections for installation on the frame.
- The overall skeleton of the garbage collector should be strong enough to prevent vibration during engine operation.
- If screws and nuts are used, the screws and nuts should be made of stainless steel.

1.1.17 SPECIAL TECHNICAL SPECIFICATIONS OF MANUAL GARBAGE COLLECTOR COMPONENTS

The manual scarper net is the same as the mechanical scarper, only its installation angle is 45? relative to the horizon.

- Garbage frame.
- Garbage collector frame should be made of steel gutter profile and have suitable supports for installing a garbage collection net and garbage collection channel on it.
- Regarding the dimensions of the frame: it can be installed in a concrete channel.
- Corrosion protection: The frame should meet the standard level after construction. It should be sandblasted and painted first with Zinc-Rich metal lining and then with suitable epoxy paint during different stages. If screws and nuts are used in the construction of the frame, they should be made of stainless steel.

1.1.18 GARBAGE COLLECTION BASKET

A metal basket with a mesh bottom, according to the following specifications, is used to collect the garbage caught by the garbage net:

- Garbage basket dimensions: suitable length for installation on the garbage collection channel.
- Steel sheet.
- Garbage collection basket should have an appropriate number of holes in the floor, 100 holes per square meter.
- Zinc-rich metal primer applied after sandblasting, followed by epoxy paint.
- Garbage collected in the basket should be emptied by the conveyor system and washed with water to the silo next to the garbage collector.

1.1.19 GARBAGE COLLECTION SYSTEM

In order to collect garbage from the net and transfer it to the basket, a four-pronged tool made of aluminum with a wooden handle is used.

1.1.20 AERATED GRAIN CATCHER UNIT

1.1.20.1 TWO-GRAIN COLLECTION UNITS

In this work, two-grain collection units are considered, which will be able to remove granular materials by using a deep aeration system. It should be planned based on emergency conditions when a unit is out of service. In this unit, the inlet part of the grain collection channel is an orifice and the outlet part has an overflow. The collected granular materials must have minimum organic materials and must have minimum moisture during disposal. The main equipment of the grain collector is as follows:

- **Diffuser Equipment:** Two mechanisms can be used for the system of collecting seeds from the bottom of the seed collection unit:

Using the Air Lift in a mobile way along the grain collection channel and continuously emptying the collected seeds from the bottom of the channel under the diffusers to the cyclone for collecting and washing the seeds.

Using a moving steamer along the seeding channel, collecting the seeds at the end of the channel and emptying them by Air Lift or a suitable pump or fixed silicone to collect and wash the seeds [56–65].

1.1.20.2 SMART CONTROL SYSTEM

Ventilation equipment includes blowers, air transfer pipes, and rotary-type diffusers. Each branch of the air pipe that enters the grain collection channel must have a valve, pressure measuring device, and a flowmeter. Each inlet pipe and its diffuser include a swing joint, upper and lower hanging pipes connecting pipes, diffuser pipe, and other accessories. A portable or fixed-hand crane should be considered for replacing the diffusers. The diffusers are installed on one side of each channel above the channel floor. The floor of the main channel has a slope towards the channel under the diffusers. The seed collection unit must include all the necessary parts and equipment to collect the seeds that settle at the bottom of the pond, and in addition to that, it guides the said materials into a funnel or container so that they can be transferred out of the pond. The devices that are placed below the maximum level of sewage to collect seeds must have corrosion-resistant and rust-proof paint. All operation steps should be automatic and only the step of discarding the seeds outside the treatment house should be done manually. The seed collection unit should be simple in terms of design and require minimum maintenance and should be considered with minimal manpower. From the point of view of starting up, the system should be usable from the control room. The lamp indicating the function should be provided in the control panel, and in addition, the on/off switch in the local panel should be considered.

1.1.20.3 SPECIFICATIONS OF THE GRAIN COLLECTOR UNIT

- **Description of the Units:**
  - The number of grain collector channels;
  - Maximum inlet flow liters per second;
  - Dimensions of each channel;
  - The length;
  - Width;
  - Liquid depth;
  - Free height;
  - Meter;
  - Time stayed in the flow as much as a possible minute;
  - Maximum grain volume taken for units;
  - Units of cubic meters per day;
  - BLOWER capacity cubic meters per minute;
  - Number of blowers.

1.1.21 FLOW MEASUREMENT UNIT

The Ultrasonic Flow Measuring (UFM) device can be installed on the pipe without a drop in pressure. These devices must be equipped with all the necessary components for current measurement and recording, such as a totalizer and stability. These devices are required for measuring the flow of incoming sewage and return sludge, and one device is for the disposal of sludge.

1.1.21.1 SPECIFICATIONS OF THE MEASUREMENT UNIT

- **Description of the Units:**
  - Diameter pipe (Mm);
  - Maximum current (L/S);
  - Number;
  - Location.

1.1.21.2 DIVISION BASIN

The wastewater exiting the granulation unit enters the flow distribution basin between the aeration tanks through a buried steel pipe. The return sludge and supernatant of the water drain pump house of the sludge dewatering system also enter the basin for mixing with raw sewage through buried pipes. The mixture of raw sewage and returned sludge and supernatant is transferred to the aeration ponds. At the entrance of the pipes inside this pond, circular steel valves will be installed to cut off and connect the incoming flow to the aeration tanks.

1.1.21.3 DIVISION BASIN EQUIPMENT

Circular steel valves with corresponding flanges and pipes are needed in this pond, as well as steel pipes. The equipment for maintaining the return sludge pipes of the water drain in this pond is also necessary.

1.1.21.4 AERATION POND

Raw sewage mixed with return sludge enters the flow division pond between the aeration ponds through buried steel pipes to enter the biological treatment stage. In total, the number of two aeration ponds is considered for the first stage of the project, and the wastewater enters the aeration ponds through orifices after passing through the selector unit. The aeration system should be provided in such a way that if necessary, each of the units can be removed from the circuit to carry out repairs and so on. The connecting pipes from the basin to the selector and round steel valve are considered inside the basin at the entrance to the connecting pipes to the selector tanks. The inlet and outlet structures of the aeration basins and their hydraulic conditions should be based on the average sewage flow plus the maximum return flow of activated sludge in the condition where one of the aeration units is out of the circuit planned.

- **The Technical Specifications and Complete Information of the Aeration:** The average flow rate is based on mass balance calculations. At the entrance of each aeration tank, a concrete tank named Selector is considered, and the exit pipes from the pond are connected to these tanks. The retention time of this pond is specified for the following situations:
  - Average discharge at the beginning of the design + sludge return discharge (mass balance calculations).
  - Average discharge at the end of the design + sludge return discharge (mass balance calculations).
  - Maximum discharge at the beginning of the design + sludge return discharge (mass balance calculations).
  - Maximum discharge at the end of the design + sludge return discharge (mass balance calculations).

It should be mentioned that aeration tanks and selector tanks will be placed in the purification circuit in the first stage, and in the second stage, aeration tanks and selector tanks will be placed.

Inside the selector tank, a depth stirrer with a horizontal or vertical flow should be used. In each selector tank, vertical depth stirrers will be installed.

1.1.21.5 AERATION TANK AND SECTOR TANK EQUIPMENT

- Surface mechanical aerator and steel or fiberglass turbine.
- Dissolved oxygen measurement system.
- MLSS measurement sensor inside the aeration tank.
- Adjustable overflow at the outlet of the aeration tank, equipped with a drive motor and the necessary gearbox
- Vertical depth stirrer with all the necessary equipment.

1.1.21.6 SPECIFICATIONS OF THE AERATION TANK FOR THE PROCESS

1.1.21.6.1 Extensive Aeration – Sewage Treatment Plant

- **Description of the Units:**
  - Number of aeration tanks (Number);
  - Volume of each pool (m^3);
  - Volume load (Kg BOD day);
  - F/m (KgBOD5/VSS.day);
  - Amount of mlss (mg/liter);

- Sludge age (Day);
- Sludge return ratio (Percentage);
- Return sludge concentration (mg/liter);
- Hydraulic dwell time (Hr);
- The surface of each aeration tank (m^2);
- Amount of production sludge (based on volatile solids) (Kg/day);
- Discharge (m^3/day);
- Oxygen required (Kg/day);
- The amount of transferable oxygen in local conditions (Kgo2/kw.hr);
- Required power (Kw);
- Energy required (Kw/1,000 m^3);
- Designed discharge (m^3/day);
- Number of aerators in each pool (device);
- Sludge discharge (m^3/day);
- Waste sludge mass (Kg/day);
- Dimensions of the aeration tank (m × m);
- The power of each fan (Kw).

The power of the aerators are modified based on the shape of the tanks and their radius of action.

1.1.21.7 THE AERATED CONCRETE BRIDGE

This part should be equipped with a fence, shelter, and rain cover to protect the engine and its equipment from the sun and rain, as well as provide access to the corridor. This equipment is also necessary for the mixer inside the selector. The aerators should be in dynamic and static balance, and their hydraulic performance should provide the required oxygen for the wastewater. In addition to the deflector plates, necessary measures should be provided to prevent sewage from splashing around the aeration basin. The shape of the vanes, the depth, diameter, and speed of the aerators should be such that the changes in the absorption depth caused by the changes in the flow and overflow regulation create uniform changes between the amount of oxygenation and the power consumption.

1.1.22 CHLORINATION UNIT

In order to ensure the necessary residence time for the disinfection of the effluent from the treatment plant, chlorine contact pond units are used, each of which will meet the maximum capacity at the end of the project.

1.1.22.1 TECHNICAL SPECIFICATIONS OF THE CHLORINATION UNIT OF THE SEWAGE TREATMENT PLANT

➢ **Description of the Units:**

- Design discharge (maximum flow) (m^3/d);
- Average sewage discharge (m^3/d);
- Maximum retention time (minute);
- Average retention time (minute);
- Dwell time at the beginning of the design (minute);
- Depth of flow (M);
- Contact channel width (M);
- The total length of the contact channel (straight length) (M);
- Number of call pools;
- The number of barriers in each pond;
- The amount of chlorine consumed in discharge is average (mg/l);
- The amount of chlorine consumed in discharge is the maximum (mg/l);
- The amount of chlorine consumed (Kg/day);
- The amount of chlorine consumed (maximum – the beginning of the plan) (Kg/day);
- The amount of chlorine consumed (average – the end of the project) (Kg/day);
- The amount of chlorine consumed (maximum – end of the plan) (Kg/day);
- The number of clarinator devices (device);
- The capacity of each clarinator (Kg/hr);
- Maximum number of chlorine capsules for one month (capsule).

1.1.22.2 CHLORINATION UNIT EQUIPMENTS

The chlorination building is separate from other buildings, and the related equipment is divided into three main parts, including chlorine supply, regulation, and injection systems. The chlorine building consists of two separate parts: the chlorine warehouse and the chlorine room. The entrance doors to them are from the outside and independent of each other. The chlorine storage has enough light, but direct sunlight should not shine on the chlorine cylinders. The chlorination building has a suitable electric fan, the key of which is in the open space. In the storage of one set of working cylinders, one set of cylinders will be ready for operation, and one set will be kept as a reserve. A one-ton cylinder will be used to supply chlorine. The cylinder is made of steel, and a ceiling crane will be used to move it. Chlorination cylinders must be equipped with reliable and tested valves that can be operated manually. Cylinder valves must have special protection to avoid any damage during transportation and storage. The cylinders are arranged horizontally so that one of the guiding pipes is in the gas phase and the other is in the liquid phase. An automatic changing device is intended to change the path of the capsules. The equipment that will be used to move the cylinders is:

1.1.22.3 ELECTRIC CRANE, LIFTING ROD

The overhead rail crane with a nominal capacity will be equipped with electric lifting and lowering. The crane should be installed in such a way that it is possible to move the chlorine capsules in the moving room, or unload or load the chlorine capsules from vehicles. The chlorinator will work based on the vacuum injection system. In this system, by creating a vacuum, the gas will move from the chlorine source and will be mixed with water in the chlorinator and applied to the injection point.

The clarinator must have at least the following equipment:

- Filtration of gas entering the clarinator.
- Vacuum injector.
- Diaphragm valve regulating vacuum differential.
- Assurance valves.
- Chlorine gas flow measurement device.
- Chlorine gas regulation valve and related electronic circuit.
- Ventilation pipe and gas inlet pipe.
- Electric heater.
- The receiver of steering signals from the remaining chlorine analyzer and the steering wheel adjusts the automatic control valve.
- Injection system for chlorine capsules.

A residual chlorine analyzer will be used for continuous measurement of residual chlorine. The residual chlorine analyzer will be located in the chlorination room. This device must be equipped with a filter and a pressure-reducing valve. The clarifier or cabinet is made of reinforced fiberglass or polyester, the flow meter is made of high-quality glass, and other parts are made of Teflon, ceramic, silver, and the like.

1.1.22.4 SAFETY EQUIPMENTS

- Chlorine gas leak detector: The leak detector will announce the presence of chlorine gas in the air. A small blower continuously directs the nearby air sample into the leak detector, and if there is chlorine gas in the air, the leak detector will trigger an alarm.
- When the alarm goes off, the blower motor must be stopped to prevent more chlorine from entering the device.
- Ventilators with suitable capacity, resistant to corrosion, can be installed on the wall and near the floor of the building.
- The ventilators should be selected in such a way that they can remove all the air in the room within 2 to 4 minutes. In order to comply with the safety considerations during the leakage of chlorine gas, before the operator enters the chlorination building, the ventilation system must work for 4–5 minutes.
- Shower and eyewash: In order to deal with possible leakage in the chlorine supply system, emergency showers with appropriate flow and pressure should be installed in a convenient and accessible place. It is also considered that a shower for the user should be near the exit doors of the eyewash.

1.1.23 SLUDGE RETURN AND DISPOSAL PUMPING STATION

The thick sludge settled in the secondary sedimentation ponds enters the telescopic tap pond installed next to each pond through a cast iron pipe. It enters through a cast iron pipe, and the sludge will be transferred from the manhole to the pumping hole of the pump house by a polyethylene pipe. The sludge discharged in the pumping station will supply the amount of return sludge required by the project, which will be carried out by the pumps installed in this station to transfer the return sludge to the pond at the entrance of the aeration tank. The rest of the secondary sludge under the title of disposal surplus sludge should also be directed from this station to the water intake units. This pump house should have the capacity to handle the average flow of incoming sewage.

1.1.23.1 QUANTITY AND QUALITY OF RETURN AND DISPOSAL SLUDGE

- Operation stage;
- Return sludge flow and disposal;
- The first stage of exploitation;

- The average inlet discharge;
- Sludge removal;
- The average inlet discharge;
- TS dry solids concentration;
- Amount of volatile solids;
- Sludge specific gravity;
- Weight of dry solids per unit volume.

In the design of sludge collection and discharge facilities, such as pipelines and pumps, the peak factor for sludge flow is considered according to the recommendations of authorities. The maximum sludge flow coefficient is determined based on the sludge treatment and dewatering facilities, as well as their storage capacity, such as secondary sedimentation tanks, sludge pumping chambers, and thickening tanks. Technical specifications of the return sludge pump station.

➢ **Description of the Units:**

- The capacity of each pump (m^3/hr)
- The number of pumps required at the beginning of the project (device)
- Diameter of thrust pipe (steel) (mm)
- Minimum speed (operation of one pump) (m/s)
- Required power of the house pump (kw)
- The minimum return sludge flow (m^3/hr)
- The maximum return sludge flow (m^3/hr)
- The number of pumps inside the chamber (device)

1.1.23.2 TECHNICAL SPECIFICATIONS OF SLUDGE DISPOSAL PUMPS INSIDE THE RETURN SLUDGE STATION

➢ **Description of the Units:**

- The average amount of sewage sludge at the beginning of the project (m^3/d);
- The average amount of sewage sludge at the end of the project (m^3/d);
- The maximum amount of sludge disposal at the end of the project (m^3/d);
- Allowed number of pumps on and off (Number per hour);
- Operation time for each pump in each work cycle (minute);

- The time of each work cycle (minute);
- Number of working cycles per day (Number);
- The working time of the house pump in one day (the watch);
- Sludge pumping capacity (station capacity) (m^3/hr);
- The number of sewage sludge pumps (device);
- The capacity of each pump (L/S);
- Diameter of thrust line (sludge disposal) (mm);
- Minimum speed in thrust line (one pump) (M/s);
- Maximum speed in the thrust line (M/s);
- Total dynamic height of pumping (TDH) (M);
- The total power of the house pump (kw).

1.1.24 SLUDGE GRAVITY THICKENER

The sludge from the secondary sedimentation units is directed to the sludge gravity thickening unit by the pumps installed in the return sludge station. Two gravity sludge thickening tank units are used for sludge thickening. The sedimentation tanks are of circular type and the sludge enters from the top of the central well, while the supernatant exits through the V-shaped overflows built into the tank environment. After thickening, the sludge is removed from the bottom of the tank to undergo the next stage of purification. The outlet structure of the thickener includes V-shaped overflows, V-Notch outlet, and an external pipe similar to sedimentation tanks. The supernatant obtained from the thickeners enters the outlet channel through the V-shaped overflows and is directed from there by a polyethylene pipe to the pump house for pumping the supernatant from the thickeners and the sludge dewatering unit to the beginning of the sewage treatment plant.

1.1.24.1 TECHNICAL SPECIFICATIONS OF THE GRAVITY THICKENER UNIT (GRAVITY THICKENER)

➢ **Description of the Units:**

i. **Referral Recommendation:**
   - Average flow rate of incoming sludge (initial design) (m^3/d).
   - Average flow rate of incoming sludge (end of design) (m^3/d).
   - The maximum amount of incoming sludge (end of design) (m^3/d).
   - Average input mass load (initial design) (Kg/day).

- Average input mass load (end of design) (Kg/day).
- Maximum input load (end of design) (Kg/day).
- Condenser tank diameter (M).
- The number of thickening tanks (Number).
- The surface of each thickening tank (m^2).
- The total surface of thickening tanks (m^2).
- Mass load entering a tank at the beginning of the design (Kg/m^2.d).
- Input mass load to two tanks at the end of the plan (Kg/m^2.d).
- Hydraulic load on a tank at the beginning of the design (M^3/m^2.d).
- Hydraulic load on two tanks at the end of the design (M^3/m^2.d).
- Amount of solid capture (%).
- Medium Supernatant output from the concentrator (m^3/d).
- The average amount of SS output in the supernatant (Kg/day).
- The average amount of BOD5 output in Supernatnat (Kg/day).
- The volume of the condensation area inside the tank (m^3).
- The height of the water at the edge of the tank (M).
- The height of the sludge thickening area in the tank (M).
- Water depth in the middle of the tank (M).

For the pumping equipment and pipelines of the sludge units, the peak coefficient is considered. However, in the calculation of the thickener dimensions, average values are used with the application of an additional value as a seasonal coefficient. The efficiency of the gravity thickener for thickening the secondary sludge without mixing with the primary sludge is not very high, and the conventional range of solids capture is equal to the calculation of the mass balance. The permissible amount of hydraulic load in the thickener unit is equal to the tank surface area and the amount of mass load on the thickeners of the surface per day. A thickening unit is placed in the circuit, and the second unit has only the role of reserve. In the second stage, both tanks will enter the circuit. It is assumed that in this case, the capacity of the thickened sludge discharge equipment is calculated based on this. It should be mentioned that the thickened sludge discharge pumps are usually plunger, positive displacement, diaphragm, or centrifugal type.

1.1.24.2 TECHNICAL SPECIFICATIONS OF THE THICKENED SLUDGE PUMP HOUSE

➢ **Description the Unit:**

- Amount of condensed sludge (first stage) mass balance (m^3/d);
- Condensed sludge mass (first stage) mass balance (Kg/day);
- Amount of condensed sludge (end of design) mass balance (m^3/d);
- Condensed sludge mass (end of design) mass balance (Kg/day);
- The maximum amount of condensed sludge (by applying a factor of 1.5) (m^3/d);
- The maximum mass of condensed sludge (Kg/day);
- The diameter of the sludge discharge pipe from each concentration unit (Mm);
- The volume of the condensation area inside the tank (m^3);
- Operation time of each pump in each work cycle (minute);
- The time of each working cycle of the pumps (minute);
- The total number of work cycles in one day;
- Working time of pumps in one day (the watch);
- Pumping capacity based on the diameter of the outlet pipe (m^3/hr);
- The number of pumps for each thickener unit (device);
- The total number of sludge discharge pumps from thickeners (device);
- The capacity of each pump (L/S);
- Fluid speed inside the pipe (m/s);
- Energy loss (M);
- Power of each pump (Kw).

1.1.25 MECHANICAL EQUIPMENT OF GRAVITY THICKENER

1. **Condensation Tank:** The thickener rotary mechanism can vary depending on the design conditions and may include vertical rods with a suitable structure as a flocculant. All equipment adjacent to the mud, such as arms or oars and harnesses, inlet shafts, screws, and U-bolts, etc., are made of stainless steel.
2. **Fixed Bridge Structure:** The fixed bridge is designed and built as a concrete or steel structure to withstand the dead and live load and the

torque of the thickening mechanism. The width of the bridge should be appropriate, and a safe fence should be installed throughout it. The floor and wall of the fence should be covered with corrugated sheet. Stairs, ladders, parapets, inspection platforms, and repairs should be considered at necessary points.

3. **Mud Removal Paddles:** Mud removal paddles are installed on rotating arms. The set of arms and paddles should be adjustable to compensate for the effects of slope and unevenness of the floor. The edge of the oars should have suitable rubber and thickness.
4. **Output Overflow:** The outlet overflow should be properly designed, adjustable with full sealing, and made of aluminum. It should be installed on the concrete wall with stainless parts and screws. The overflow must be adjustable in terms of installation height.

1.1.25.1 TECHNICAL SPECIFICATIONS OF SEWAGE ELECTRICAL INSTALLATIONS

1.1.25.1.1 Medium Pressure Panels

- **Plan Conditions:** All equipment must be designed in such a way that it can withstand all mechanical, electrical, and other stresses or destructive effects that may occur in practice. All equipment must be installed inside the building.

1.1.25.1.2 General Technical Specifications

- **Insulation Degree:**
  - Rated voltage;
  - The maximum voltage of the system;
  - The value of the impulse voltage;
  - Frequency.

1.1.26 THE MAIN DETAILS OF THE CELL BUILDING

The cells must be metal and they must also be able to be installed on the channel. The location of the supports and the installation location of the cables should be created in the end part of each cell. The manufacturer must fully respect the allowed distances between the copper busbars and the cables according to the standards used. The specifications and installation method of the internal equipment of the panels should be designed in such a way that it respects the reliability coefficients and the equipment that is necessary for the operation and control of the automatic switching. In order to strengthen the cell and install supports for insulators and keys, each cell must have a metal frame. The internal equipment of the cell should have such good quality that maintenance and repair problems can be reduced to the minimum possible. The internal equipment of the panel should be installed in such a way that there is no vibration during operation and the control equipment should be protected against noise. All levers and metal parts must be made of corrosion-resistant metals. Each cell must have a separate door. The doors should move easily on the hinges and the hinges should bear the weight of the door well. Metal material and hinges should be galvanized steel or phosphor bronze. Each cell must have a separate lock and also the necessary interlock between the door and the equipment as well as the generator and the sectioner. It is mandatory to observe the standard distance between the moving parts of the equipment and the panel body [66–72].

1.1.26.1 THE TIRES

Copper should be used to make the main bus bars of the cell. The main busbars of the panels must be placed on the standard insulators in the panel. In order to withstand the rated current of the panel and the thermal and dynamic forces caused by the short circuit, the busbars must have appropriate dimensions. Each busbar must be of a standard and heat-resistant color, with the following colors painted:

- R phase red;
- Yellow S phase;
- Blue T phase.

The distance between the busbars in each option must be according to the standard. To connect the cells to the ground, a ground busbar with suitable dimensions must be installed on a standard support insulator. The metal body of the cells must be connected to this bus. The connection of cells and tires should be free of grease and paint. The ground cover should be painted with a light gray color. In order to connect the earth wire of the cables, a suitable hole should be considered on the ground bus. The metal parts of sectioners and degenerators must be firmly connected to this bus.

1.1.26.2 SCREWS, NUTS, AND WASHERS

All the screws, nuts, and washers used in the panels must be of high quality, made of stainless steel or galvanized. The dimensions of screws, nuts, and washers should be in the metric system. It is necessary to use a spring washer under all nuts.

1.1.26.3 EXPLOSION VALVE

In order to prevent the board from exploding during the event of a short circuit and other defects, the manufacturer must install a special valve. This valve should be designed in such a way that it works well in the event of a short circuit and directs the effects of the short circuit to the outside.

1.1.26.4 CERAMIC SUPPORTS, INSULATORS, HOUSINGS, AND BUSHINGS

Insulators should be made of first-grade standard porcelain or similar materials and should be free of cracks, seams, or damage, and glazed in all different directions. The insulation tests of internal and external substations of systems with a nominal voltage. Their housings must be designed in such a way that they can withstand electric arc or electric discharge without damaging the seat or other main parts. In addition, they must be designed in such a way that they can withstand all reactions caused by shock and gravity. All parts of the enclosures, except the gaskets, must be made of materials such as metal or glazed porcelain that do not absorb moisture at all. Necessary measures must be taken to compensate for the expansion of conductors and current-carrying parts.

1.1.26.5 COLORING

All the surfaces of the cells and other paintable parts must first be smoothed, and after washing, degreasing, phosphating, and other necessary operations, they should be painted by the electrostatic method with a thickness of at least 80 microns and in two layers, using standard paint.

1.1.26.6 EXPERIMENTS

The following tests on the cells must provide:

- Testing the degree of protection.
- Testing the mechanical resistance and the amount of pressure and resistance of moving parts and contacts.
- Measuring the voltage drop on the phases.
- Temperature test under long-term operating conditions.
- Insulation test of primary and secondary circuits.
- Measurement of insulation resistance.
- Thermal and dynamic stability test.
- Testing the capacity of cells and keys.
- Testing the action of open electric arc in cells.
- Test of protective coatings.

1.1.26.7 GENERATORS

The work that is done according to these specifications includes the design and preparation of all drawings and technical documents, manufacturing, testing, delivery, installation, full commissioning, and guarantee of generators. The generators must be equipped with all the necessary equipment, including steering devices with opening and closing mechanisms, cable trays, and necessary wiring, so that connecting the external conductors to the terminal box is sufficient to use the generator.

1.1.26.8 TECHNICAL SPECIFICATIONS

Generators of the vacuum type (according to the diagnosis of the monitoring device) are selected and fixed (or sliding based on the diagnosis of the monitoring device) for installation inside the building.

1.1.26.9 DESIGN AND PERFORMANCE CONDITIONS

- Number of poles;
- Nominal system voltage (kV);
- Maximum system voltage (kV);
- The effective value of a shock with the grid frequency for a period of one minute (kV);
- Nominal frequency (Hz).

1.1.26.10 CERAMIC SUPPORTS, INSULATORS, THEIR HOUSINGS, AND BUSHINGS

Insulators must be made of high-quality porcelain or similar materials. They should be free of cracks, seams, or damage, and have a glazed surface in all directions. The insulation tests of internal and external posts of systems with a nominal voltage must comply. Ceramic supports, their enclosures, etc., must be designed in such a way that they can withstand electric arcs and electric sparks without damaging their seats or other main parts. In addition, they must be designed in such a way that they can withstand all the reactions caused by shock and gravity. Ceramic enclosures must be designed in such a way that no additional stress is applied to any of its parts due to temperature changes. All parts of the enclosures, except the cassettes, must be made of materials such as metal or glazed porcelain, which do not absorb moisture at all. Necessary measures must be taken to compensate for the expansion of conductors and current-carrying parts. All the sealing parts must be properly sealed to withstand the pressure caused by the temperature changes caused by the normal operation of the switch and also the ambient temperature changes without leaking or moisture absorption [73].

1.1.26.11 CONNECTING AND DISCONNECTING DEVICES AND MECHANISMS

Generators must be equipped with a disconnecting and connecting mechanism. Disconnecting and connecting generators will be done manually using spring charging. It is not possible to operate the key while charging the spring. When the spring is charged and released, it should be indicated on the key by a mark. The handle for spring charging should be designed in such a way that when the said handle is connected, it is not automatically rotated or thrown around. Also, the disconnection and connection sign of the key with two different colors should indicate the state of disconnection and connection of the main contact. In addition, the mark of charging and discharging of the spring must be clear on the key.

1.1.26.12 GENERATORS CONTROL

Connecting and disconnecting the generators will usually be done. However, the necessary arrangements should be considered for remote control status, emergency situations, or for testing the system and operation of the protective circuits and relays in the switchboards. The necessary interlock should be provided in connection with opening the cell door and disconnecting the generator through the corresponding trip coil.

1.1.26.13 KEY STATUS

Each degenerator must be equipped with four sets of open and closed auxiliary contacts, except for the contacts that are required for the operation of the mechanism. These contacts will be used to indicate the status of the key.

1.1.26.14 CURRENT CARRYING PARTS

Current-carrying parts must be designed and built in such a way as to meet all specified requirements such as rated current, short-term rated current, rated breaking capacity, rated connecting current, mechanical resistance, and resistance to corrosion. Contacts and terminals should be made of materials that have low resistance to passing current. The temperature increase of each part of the degenerators should not exceed the permissible limits of temperature increase. The contacts should be replaceable and their contact surfaces with each other should be stable and complete, and they should not require unnecessary and additional repairs. Each degenerator must be equipped with the necessary cut-off coils to apply the considered protections in each part. The signs indicating the state of the key in open and closed states, as well as the key lock, must be installed in front of the panel door. The control wiring should be made by copper wire with P insulation. All the wires must be connected to the terminals.

1.1.26.15 SPARE PARTS

The spare parts needed for each generator will be as follows:

- Auxiliary contacts;
- Springs, bobbins, and other small parts.

1.1.26.16 GUARANTEED ITEMS AND CONDITIONS OF NON-ACCEPTANCE OF GENERATORS

The breaking capacity, operation time, and amount of continuous and alternating current consumption must be guaranteed. Additionally, generators will not be accepted if any of the following conditions are met:

- The cutting power is less than the standard value and the operation time is more.

- The amount of constant current consumption exceeds the standard value.

1.1.26.17 TESTS

All the tests must be done are as the following:

1. **Routine Tests:** Normal tests should be performed on the generator and are as described below.
2. **Generator Usage Tests:** "POWER FREQUENCY" tests.
3. Measuring the resistance of the main circuits.

1.1.26.18 SAMPLE TESTS

The certificate of sample tests should be provided to the generator as described below:

- Mechanical tests;
- Temperature increase experiments;
- Impulse voltage tests;
- Breaking capacity tests;
- Connection capacity tests;
- Short-term current tests;
- Transient recovery voltage tests.

1.1.26.19 GROUND CONNECTION SWITCH

The work includes the design, preparation, and presentation of all engineering documents, construction, testing, transportation, installation, and guaranteed commissioning of grounding switches. The mentioned keys are complete and include all necessary accessories. These keys are installed inside the panel.

1.1.26.20 DESIGN AND OPERATION SPECIFICATIONS

➢ Technical Specification:

  - Number according to the map and engineering documents.
  - Nominal system voltage (kV).
  - Maximum system voltage (kV).
  - Frequency (Hz).
  - The degree of insulation is the same as the specifications.

- **Main Construction Details:** The earth connection keys are installed inside the panels in a way that they can be closed and establish the earth connection when the main tab of the automatic keys is cut off. The mechanical connection of earthing switches with automatic switches will prevent the switches from operating and being connected to the main busbar as long as the earthing switch is closed. The opening and closing of these keys in manual mode will be done by spring charging. The operation of the key is complete and under any conditions, and it will be in such a way that no energy remains in the actuation spring after the operation. In other words, the top key has two independent positions: "open" and "closed." The "open" and "closed" positions of the key should be marked and colored. The handle corresponding to the operating mechanism of the key must be placed in such a way that it has a sufficient distance from the busbar and the cables. The mentioned key should also be protected against corrosion, and its metal parts should be made of galvanized steel or chrome.
- **Measuring and Protection Devices:** The measuring, indicating, and protective devices installed in the relevant boards should be as follows:
    - Green indicator to show the open state.
    - Red indicator symbol to show the closed state of the key.
    - Free current and primary or secondary relays.
    - Grounding relays.
    - Protection related to the increase in temperature of the transformer oil and other signals sent from the transformer.
- **Construction and Main Characteristics of Protective Relays:** Protective relays must have an anti-dust cover and be suitable for flush installation in the panel. The relay cover must be equipped with a target so that it can be returned to its original position from the front without removing the cover. The test push buttons should also allow for testing the relays without removing the relay cover.
- **High Current Protection:** For high current protection, three-phase adjustable delay high current relays should be used to protect against short circuits and to protect the design transformer. These relays must have the following specifications:

1. **Adjustment Range:** The cut-off time delay can be adjusted based on the amount of overload.
2. **Earth Connection Protection:** A high voltage relay is used to detect ground connections. This relay should be connected to the open triangle winding of the voltage transformer. The relay mentioned above should cut off the power supply switch in the event of an error.

1.1.26.21 MEASURING EQUIPMENT (CURRENT TRANSFORMER, VOLTAGE, ACTIVE AND REACTIVE POWER AND OTHER RELEVANT EQUIPMENT)

All measuring devices must be dust-proof and suitable for parallel installation in the panel, and these devices must be of a tested type, complete with all necessary accessories such as measuring transformers and indicators.

1.1.26.22 EXPERIMENTS

- The tests related to the measurement of the accuracy class.
- The performance characteristics of the measuring devices.
- The relevant current and voltage transformers.

In order to perform the tests, three-phase current and voltage should be applied to the input terminals of the devices with a suitable phase difference (similar to the actual working conditions) in order to determine the direction of movement of the moving element of the respective device and to determine and measure the electrical quantities given to the device. The performance of protective relays should be tested. For these tests, appropriate current and voltage should be applied to the input terminals of the relays. Three-phase potential and current should be applied to the three-phase protective relays that have directional performance characteristics for conducting tests.

1.1.27 CABLES

The work that is done according to these technical specifications includes: designing and preparing all necessary documents, constructing and testing, transporting, and installing cables. The cables will be single phase and have a suitable cross section.

- **Technical Specifications:**
  - The number of strands.
  - Type of copper conductor.
  - Polyethylene insulation type.
  - Nominal voltage (kV).
  - Maximum system voltage (kV).
  - Cross section according to maps and documents.
  - Frequency (Hz).
  - Maximum nominal heat for insulation for nominal current for short connection current.

- **Construction Specifications:**
  1. **Conductors:** These should be of the best type of annealed copper, and they should be made of twisted wires. The conductors should be able to withstand the heat increase allowed for passing the nominal current. The specifications of the conductors must be in accordance with a valid standard. Also, the conductors must be able to withstand short circuit currents and mechanical and thermal stresses according to the standard and design conditions.
  2. **Insulators:** These are made of a dielectric with a wide spectrum of the best type of material, which is usually molecularly pure polyethylene. The insulation will be such that it can withstand the maximum working voltage of the system, as mentioned above, during continuous operation, without having a weak point. Also, the insulation should be in a position to minimize the insulation loss in the cable. The insulation thickness should not be less than what is mentioned in the valid standard.
  3. **Metal Mesh:** The metal mesh in the cable consists of one or more layers of metal tape wires that are placed around the cable in a suitable layer depending on the cable design. This net should be free of any cracks, holes, or impurities.
  4. **External Non-Metallic Coating:** All cables will have an external non-metallic coating. This cover is made of heat-resistant plastic material (thermoplastic) or other similar material. This cover must have enough resistance for the permissible cable heat according to the valid standard.

1.1.28 EXPERIMENTS

All cables must be tested in accordance with acceptable valid standards:

1. **Normal Tests:** Measuring the electrical resistance of the cable conductor.
    - Partial discharge tests;
    - Resistance tests against maximum voltage.

2. **Electrical Tests:** The certificates for the following tests must be provided. These are the tests that have already been done in the laboratory.
    - Partial discharge test;
    - Bending test, plus partial discharge;
    - Measuring tangent angle as a function of voltage and cable capacitance;
    - Thermal cycle test in addition to partial discharge test;
    - Instantaneous voltage resistance test, along with Power Frequency voltage test;
    - AC high voltage test.

3. **Non-Electrical Tests:**
    - Insulation thickness measurement.
    - Measuring the thickness of the non-metallic coating on the cable.
    - Determining the mechanical characteristics of insulation before and after use.
    - Water absorption test in insulation.
    - Test against flame.
    - Special bending test.

4. **Electrical Tests After Installation:** The transformers will be installed inside and outside the building, in three phases, with a cooling system of natural circulation of oil and air. Connections on the high and low voltage sides are made by cable. All the necessary calculations for the design and preparation of the required engineering documents, construction, and safe operation of transformers have been completed.

5. **Nominal Specifications:** In addition to the above, the manufacturer will ensure that all equipment in this line complies with the following specifications and any other required specifications.
6. **Nominal Permanent Capacity:** The number and permanent capacity of transformers are confirmed according to technical drawings and engineering documents.
   - Nominal voltage;
   - Initial voltage: kilovolts;
   - Secondary voltage volts;
   - The initial voltage adjustment;
   - Frequency: Hz;
   - Wire connections Dyn;
   - The zero point of the secondary circuit will reach directly abroad;

The degree of insulation resistance will be as follows.

i. **Medium Pressure Side:**
   - Rated voltage kV;
   - The maximum network voltage kV;
   - Impulse voltage kV;
   - The maximum test voltage Hz and for one minute kV.

ii. **Secondary Side:**
   - Rated voltage;
   - Impulse voltage;
   - Short circuit impedance: according to the specifications of the table;
   - Maximum sound level: according to the table.

iii. **Maximum Allowed Load Capacity:** The maximum permissible temperature of the transformer, according to the valid standard, is as follows:

1.1.27.1 THE MAIN DETAILS OF THE CONSTRUCTION OF TRANSFORMERS

The transformers are oil-type, three-phase, with a cooling system of natural circulation of oil and air. The transformer is sealed with the body and oil is pumped into a tank that can absorb the volume of oil caused by its heating and cooling.

1. **The Body of Transformers:** The body of the transformers is of a type where the strong and weak pressure insulators are connected to it by a flange, and a special grounding and crane ring is installed on it. The aforementioned flanges are strong and weak in order to facilitate installation and show the pressure points. A special valve is provided on the transformer body to drain the oil and take a sample of the oil for chemical and electrical testing. The said body will have the power to withstand pressure.
2. **Cooling System:** The transformer is equipped with an oil cooling radiator. The oil can have a natural circulation that is caused by the change of the specific weight of hot and cold oil inside the radiators. The mentioned radiators will be cooled by natural air circulation.
3. **Voltage Change Key:** The switch to change the voltage in the direction of the strong pressure of the transformer is of a rotary type and can only work when the transformer is unloaded. The device for moving the voltage change switch is installed on the transformer body and will be equipped with a special program to prevent unauthorized access. In the construction of the transformer, it should be observed in a way that solves all the problems related to the way of communication and prevention of damage caused by short circuit.
4. **Accessories:** The transformer will include the following accessories:
   - Indicates the oil level in the tank.
   - Mercury thermometer with metal sheath.
   - Relay to protect the transformer from possible dangers.
   - Temperature relay
   - Ground connection point.
   - Stuck
   - Place for pouring oil.
   - Oil drain valve.
   - Errors and signals in the transformer can be included and displayed in the control system by placing appropriate equipment inside the relevant panels.
5. **Corrosion Protection:** The transformer will be painted with special paint. The metal parts of the transformer must be protected against

corrosion. For this purpose, after cleaning the oil from the metal parts, they should be painted with several layers of industrial paint.

6. **Spare Parts:** The following spare parts will be provided and delivered for the transformer:
    - A number of insulators of each type.
    - A series of washers for transformers.
    - Transformer oil

1.1.27.2 EXPERIMENTS

The factory tests that must be done at the manufacturer's factory on the transformer will be as follows:

- Test oil;
- Measurement of insulation resistance;
- Measuring the resistance of conductors;
- Measurement of insulation losses;
- Control of communication of coils;
- Measurement of no-load current and drop;
- Measurement of primary and secondary voltage ratio;
- Insulation test by alternating voltage increase;
- Phase connection control;
- Controlling the slope of the connection pipe between the oil tank and the transformer.

1.1.27.3 DELIVERY

Transformers must be transported in an oiled form and closed in such a way that no damage can be caused to them during transportation, whether it is dry or on the ground.

- Installation location of transformers and its accessories.
- Foundation details.
- Loads applied to the foundation.

The height of the main components of the transformer, such as the high and low voltage terminals, is indicative, and so on.

- Wiring plan for protection, control, and measurement.

- Magnetic saturation curve.
- Details of making insulators.
- Operation and maintenance instructions.
- The catalog and technical information are related to each of the systems and tools.
- Technical information about oil properties and tests.
- Low pressure equipment.
- Weak pressure panels.

1.1.27.4 JOB DESCRIPTION

The work that is done according to these specifications includes the design and preparation of the necessary documents, construction, final testing, and the delivery and guarantee of the alternating current panels:

- Voltage.
- The nominal working voltage volts, Hz, four wires, alternating current.
- Normal current and short circuit.
- All devices must be able to operate continuously for the thermal relay to work properly. Additionally, the devices must withstand the thermal and dynamic short-circuit currents specified in the technical documents.
- Local protection
- All alternating current systems must be locally and selectively protected. This protection should be done against the fault current and high load. In this regard, it is necessary to properly set the timing and the settings of the protection relays used.

These main panels will be supplied from the main transformers. In general, the following panels are included in the plan:

- Several main electrical panel devices.
- Several power supply panels for each unit.
- A few devices of low pressure panel for domestic use.

Each of the panels is made up of different components, which will be discussed below:

1. **The Main Electrical Panels:** The main panels at least include the main panels of normal electricity and emergency electricity.

2. **Power Board of Each Unit:** The power supply board of the units should have at least the following items:

    - An input cell to feed the panel.
    - Automatic key.
    - Keys related to the automatic key.
    - Keys related to the technical documents.

    According to the diagnosis of the monitoring device, all electric motors that have the ability to send error signals (such as coil heating, electric motor errors, etc.) should have their errors and other problems appear on the board created in the control room system. For all the electric motors that have the ability to send error signals (such as overheating, electric motor stalling, etc.), it is necessary to add relays, contacts, and other necessary equipment on the panel. The errors and other items of the control room should be shown in the system in the control room. The operation time of all the electric motors in the system should be recorded and visible [74].

3. **Local Boards:** For the use of different units, the control of each of the electropumps should be provided locally. These panels should have at least a selector key for local/remote/off modes, signal lights, and Start/Stop buttons.

    - The building of panels.
    - Minimum technical specifications.
    - The signs must be of metal type.
    - The different parts of the panels should be large enough to withstand all the accidental stresses caused by transportation and installation without being damaged, and withstand the dynamic force of the short-circuit current.
    - The arrangement of the devices inside the panel should be such that they can be inspected and repaired as easily as possible.
    - Ventilation should be provided in all metal panels. Ventilators should be made in such a way that the wires and devices inside the panel are not damaged by water leakage and dripping. These ventilations should have anti-dust filters and nets to prevent dust and insects from entering the panel (the panels with high IP are without ventilation).
    - Panels should be made in such a way that they can be connected without opening the panel. Panel doors should be made of a steel sheet.

- In the power panels, while using air circulation ventilation through the regulator, the anti-heat heating system is also used through thermostat and all panels.
- All the bodies and cells must match, painted.
- The heater inside the panel should work based on the internal sensor that is installed in the panel.

1.1.28 METAL BUILDING

The panels are made of steel sheets and are of modular type. The boards should be formed in such a way that they provide a free environment. The skeleton of the panels should be used to hold the devices and pipes that are installed behind the screen and in front of the panel, and the structure of the panel should be such that it can be connected to the foundation with screws and nuts. Bolts and nuts are made of corrosion-resistant alloy. Locks and hinges must have gaskets or similar.

1.1.28.1 ENOUGH TIMES

Busbars are made of copper and are kept by insulators. Busbars and connecting rings must be screwed together. The connecting surface of the bus bars should be silver or silver-like metal. The ground load bus must be isolated from the panel body and connected to the ground system terminal with a detachable connector.

1.1.28.2 WIRING AND CONNECTIONS

The internal connection of panels with an excellent appearance and with plastic insulation with sufficient cross-sectional area is done. Each wire must withstand short-circuit current and overload without increasing its temperature. The wires must be firmly connected to the terminal so that vibrations and other complications do not cause them to break. The terminals must be sufficient so that each input and output conductor is connected to a terminal. In internal wiring, each one must be connected to one terminal, but in electrical devices, if the terminal visibility is sufficient and it is desirable that a maximum of two wires can be connected to one terminal. Complete isolation must exist between terminals with different voltages and different functions. Terminals should be considered for the secondary direction of the current transformer. These terminals should have means to make it easy to short the secondary of the transformer. In addition, the mentioned terminals must be designed in a way that can easily accommodate testing devices that have plug-in type connectors. All terminals must have numbers, and wiring must also have serial numbers.

1.1.28.3 WEAK PRESSURE AUTOMATIC KEYS

The low-pressure automatic switches feeding the panels have high breaking power and are able to be manually and automatically disconnected and connected. The switches feeding the motors can be connected manually, but in the event of a short circuit or overload, they will automatically connect the electric circuit. The minimum insulation class of main switches and compact switches contacts must be connected and disconnected quickly. A mechanical indicator that shows the closed and open state of the key should be included for each key. The protection circuit should include a thermal current relay and an electromagnetic relay for short circuit. The keys must have auxiliary contacts to operate the indicating devices when the key is cut by the thermal and magnetic relay.

1.1.28.4 OVERLOAD CONTACTORS AND RELAYS

Overload contactors and relays should be adjusted according to the rated current and starting current of their feeder circuit. Manufacturers of metal-free contactors and relays should be selected from well-known and reliable manufacturers. In choosing contactors and bimetals, maximum efforts should be made to change them.

1.1.28.5 ENGINE STARTER

Starting the electric motors according to the drawings and technical documents is selected as soft starter, star and triangle, or direct (with the opinion and diagnosis of the monitoring device).

1.1.28.6 ELECTROMOTOR SPEED CONTROL SYSTEM (VSD)

Since in the current design, the electric motor speed control system (VSD) is used to change the comfortable air to the desired value, this system must have the following technical specifications:

1.1.28.7 FRENCH METHOD (UIF)

The selected speed control system should create the least and weakest power network, and in other words, it should create the least harmonic in the network. The current and voltage of the electric motor in different revolutions of the motor should be sinusoidal and with minimum drive harmonics. It should provide the minimum torque fluctuations (Torque Pulsation). The drive should provide the user with references on the load that are valid in similar applications.

The structure of the drive should be modular and allow the user to develop their own programs using powerful internal software. Protect the electric motor against current and voltage fluctuations. Build it close to the motor so that there is no need to use a reactive power compensator (rectifying capacitors). There should be no limit applied by the manufacturer for the length of the output cable from the drive to the electric motor. The drive does not need to use a filter, even if it is possible, in order to make the electric motor compatible with the drive box. The motors must have greaseable shafts with a grease chamber. The starting method for motors and their rated current must be limited to the full load current. The triangle was (automatic) and its starting current should be limited to the full load current.

The motor circuit should be equipped with aids such as indicator lights, on and off buttons, an operation clock, an ammeter, auxiliary relays, and time relays.

The key for manual-off-automatic status or local control, remote control, and protection devices will be as follows.

- Motor and cable protection against short circuit current.
- Motor protection against short circuit current.
- Motor protection against load.
- Motor protection against voltage shortage or prevention of single-phase or two-phase motor operation.

The electric motors used for the Moore surface fans must be designed to be started by the remote control method (VSD) so that "the bearings and the cooling fan do not have a problem in their operation" in terms of insulation.

1.1.29 CURRENT TRANSFORMERS

Current transformers generally have an accuracy class and should be considered according to drawings and technical documents wherever necessary. The sizes of current transformers must match the power of the measuring and protection devices they are connected to.

1.1.29.1 MEASURING DEVICES

Voltmeters and ammeters should have a device for zero adjustment and the degree of the device screen. These devices should be prominently displayed on the boards. Their pages are white, but the numbers are written in black. There is no measurement error in these devices. The ammeter for the motor must have a compact rating at the end to control the starting current. The power of measurement and current of measuring devices and relays should be equal to the output of transformers [75].

1.1.29.2 DESIGN CONSIDERATIONS

- **Calculations According to the Size and Power of the Keys:** Standards, calculations, and maintaining position and selective for all systems. DC, AC.
- **Maps:**
  - Diagram of all wiring.
  - Diagram of all connections and internal connections.
  - List of all components.
- Equipment Catalog: All cables with mention of type and cross section for internal connections (including calculations of current, loss, latency, and short connection).
- Excluding Plans and Installation of Signs:

Installation and Foundation Plans:

As Built Maps

- Experiments;
- Factory tests;
- Factory tests include at least the following steps.

➢ **General Inspection and According to Construction Plans:**

- **Dielectric Test for Each Device:**
  - Output measurement and measurement problems and related transformers.
  - Tests on how to protect against excess current and short circuit.
- **Spare Parts:** The list of spare parts is as follows.
  - An automatic key of any type.
  - A complete set of thermal and magnetic key relays of all types.
  - A contactor of any type.
  - A motor protection key of any type.
  - A soft launcher of any kind.
  - A remote control device of any kind.
  - A voltmeter of any kind.
  - An ammeter of any type.
  - 100% of the lamps show.

Around 10% of all other equipment.

- Weak pressure cables.
- Job description.

The work that needs to be done in this problem includes designing and preparing documents, manufacturing, supplying, delivering, testing, and guaranteeing low-pressure cables and other necessary equipment in the cable laying project.

1.1.29.3 TECHNICAL TYPE

Cables have different conductors. These cables have double insulation. Each individual conductor has isolation. The cables have a universal sheath. Cables are used to distribute energy at low-pressure voltage.

1. **Specifications and Nominal Values:**
    - Nominal voltage in the power frequency between the conductor and the ground kV.
    - Nominal voltage between the conductors at a power frequency kV.

2. **Conductors:** Cable conductors must be of high quality and pure copper.
3. **Insulation:** The insulation of auxiliary voltage cables must be made of solid polyvinyl chloride insulation according to the valid standard.
4. **Cable Trays and Ladders:** Cable trays and ladders are made of bendable steel plates and are hot-dip galvanized or PVC-coated, with the necessary bands and manufacturable edges. The thickness of the sheet and the number and distances of the supports should be such that under the influence of the above-mentioned loads, visible curvature should not be created in the tray. The cable tray is connected to the main ceiling on metal supports with the ability to change and adjust the height, position, and end of the arm of the supports. They are made similarly. The sections and the straight path of the cable tray should be such that the bend created at the end of the straight path does not exceed the permissible bending of the cable. In the vertical direction, the cable is fixed to the ladder. The package should be suitable for the size of the cable so as not to damage the cable. All cable ladders have a cover.

1.1.29.4 EXPERIMENTS

- All tests should be according to IEC 502 standard.
- Normal tests
- Conductor resistance
- Voltage test
- Sizes

1. **Sample Tests:**
    - The manufacturer must certify.
    - Electric

- Measurement of insulation resistance at ambient temperature.
- At maximum temperature.
- Voltage test for four hours.

Overcurrent and short-circuit current tolerance test under existing conditions:

- Non-electrical
- Thickness measurement.
- Measurement of mechanical properties.
- Measurement of thermoplastic properties.
- Thermal shock test.
- Water absorption test.

2. **Earth System:** The transformer posts will be shared with the ground network for the design of a complete ground network. The mentioned ground network also includes hydraulic metal parts such as water pipes and the metal skeleton of the building. It can be measured.
3. **Design Specifications:**
    - The maximum earth connection current.
    - Resistance for medium lands: one should pay attention to the real characteristics of the ground system that will be measured by the test.
    - Maximum connection time for the ground: The maximum time for ground connection, considering the allowed time of short connection in the transformer and the operation time of the relays.
    - Voltage and pitch: The contractor will design the maximum contact voltage and pitch of the ground network in such a way that the changes of the contact studies conducted on the ground surface are in accordance with the criteria recommended in the valid standard.
    - Earth network resistance.
    - The resistance of the ground network.
    - The ground grid is made up of vertical electrodes and horizontal strips. The components of the ground network must be specified by the calculations, maps, and documents provided by the contractor. This general network will be for equipment ground system (service), protective ground, arrester, etc.

- The earth system must be applied in a complete way everywhere and in the cable trays and in the connected parts, it must be done specifically through the necessary jumpers.

4. **Specifications of the Materials:** The manufacturer will provide all the electrodes, tapes, connections, and other accessories needed for the ground network.

   Consumables, communication conductors, communication devices, and ground network, as well as other related parts, will be completely made of copper and of the thermal welding type.

5. **Power Factor Correction System:**

   i. **Technical Specifications:** The power factor of the system must be corrected by a group of capacitors, which are controlled by the automatic regulator.

   ii. **Design and Efficiency:** The continuous rated capacity of each capacitor should be determined based on calculations, technical drawings, and related documents in order to completely correct the power factor and eliminate network harmonics.

      - Nominal voltage of low pressure capacitors.
      - Nominal capacity.
      - Construction standards
      - Connection: Triangle
      - Cooling: with natural air

1.1.30 THE MAIN DETAILS OF THE CAPACITOR HOUSING

Each capacitor unit will have a metal case. This container is made of stainless steel and should be strong enough for transportation. The interior surface should be finished according to international methods and performance standards. Sufficient attention should be paid to tightening and closing the insulators. The internal and external parts of the capacitors must maintain their strength during long-term operation. The network is insulated.

1. **Spare Parts:** The capacitor system spare parts are as follows:
   - Contactor;
   - Fuses;
   - Discharge resistance.

2. **Protective Equipment:** The capacitor units must withstand all the voltage whose power value is equal to the nominal voltage of the capacitor terminals. These capacitors must withstand a current that does not exceed the breakdown value of the current passing through the capacitor at the nominal sinusoidal voltage and nominal frequency (without considering the transient states), and the capacitor assembly must withstand the voltage and current caused by harmonics. Capacitor sets have been made to reduce harmonics. The capacitor elements must be assembled with suitable protection fuses so that defective elements can be immediately separated by this fuse.
3. **Discharge:** The capacitors must be discharged through external resistors in such a way that their residual voltage decreases after disconnecting the capacitor from the power source.
4. **Connections:** Each set of capacitors must be connected in a triangle. The external connection of the capacitor set to the electric power source will be done by the cable and through the cable glands placed on the cover of the terminals.
5. **Specification Plate and Marking:** Each capacitor unit must be equipped with a nameplate. This nameplate must be installed in a visible place and include the minimum items accepted in the valid standard.
6. **Corrosion Protection:** For proper operation, the capacitor units must be painted.

1.1.30.1 EXPERIMENTS

1. **Routine Tests:**
    - Capacity.
    - Capacitor drops.
    - Direct voltage test between terminals.
    - AC voltage test between terminals.
    - Test AC voltage between terminals and housing.
    - Tests between terminals and ground for each capacitor assembly.
    - Necessary tests for the optimal functioning of each capacitor set, together with the regulating unit.

2. **Sample Tests:** Capacitor drops at different temperatures.
3. **Thermal Stability Test:**
   - AC voltage test between the terminals and the housing.
   - Impulse voltage test between the terminals and the housing.
4. **Discharge test.**
5. **Ionization test.**
   - Tests at the Installation Site:
     - General visit.
     - Visiting electrical and field communications.
     - Insulation test for each capacitor.
     - Testing the performance of each set of capacitors manually and automatically.

1.1.31 STREET LIGHTS

- The bases of street lights and floodlights must be of the type, shape, and measurements specified in the specifications and maps, and they should be able to withstand wind vibrations, among other factors.
- The street lamps and floodlights must comply with the given specifications and should not be vulnerable to wind, rain, or dust. In explosive areas, they must be explosion-proof and have official safety documents.
- Street lights and lighting towers should be easily accessible for lamp replacement, and the floodlight masts should have a ladder and work area above if necessary.
- The connection box of the lights should be installed at the bottom and on the base plate. The main cable entry and exit, as well as the branching cable exit to the lamp, should have the necessary equipment such as fuses and cable guides.
- The foundation installation in the ground and the cable entry and exit should be done according to the plan and specifications.
- The basket of light towers should be mobile, and for the movement of its head, the gearbox and electric winch should be considered.
- Street lights and floodlights must have correct and complete light efficiency curves according to the requirements of the location and maps, and for the map to be usable.

1.1.32 *FIRE EXTINGUISHING EQUIPMENT*

In every treatment plant, the presence of flammable gases in sludge facilities and chemical storage facilities, maintenance modification workshop, and administrative and personnel areas, as well as the use of disinfection facilities and the presence of flammable materials in different units, should be considered when installing fire extinguishing facilities. These facilities should be compatible with the environment (danger of electric shock) of flammable materials and chemical fumes.

1. **Diesel Generator:** To provide electricity to the necessary equipment during a power outage, emergency diesel generator devices should be provided according to the design specifications, calculation notebooks, and related maps.
2. **Equipment Brand:** The diesel generator is designed for working in emergency conditions, and it is necessary to have all related equipment and accessories such as a heater, exhaust battery, etc. The equipment must be new and covered in fabric.
3. **Automatic Startup:** The generator panel must have all the necessary protections and be able to start and work automatically so that all other systems (fuel starter, heater, and related items) work automatically. It is possible to start the system manually. All signal lamp control lines and specification plates can be seen on the board. The diesel generator should be automatically integrated into the system and turned on and off by the operator in the control room, and its status should be displayed.
4. **Diesel Engine:** The diesel engine must have at least a lubrication system, fuel injection, coolant, start-up, speed regulation, suction filter, control and measurement equipment panel, and precision instruments. It should also have an exhaust with a muffler and a manual and automatic fuel transfer system.
5. **Generator:** A three-phase synchronous generator with brushless (self-excited) copper coils, anti-moisture penetration, with class F insulation and self-cooling, should be used. It should have a temperature tolerance and can be used in areas with high humidity. The generator should be able to provide the rated power with its speed and have output voltage regulation independent of the power coefficient, with a low harmonic level. It should be protected against

electrical opposition, load, heat increase in the stator and bearings. Additionally, it should have a distance meter, temperature detector, and anti-heat heater.

6. **Control System:** The planned control device for the purification house is in the form of manual, local, and automatic control, along with central supervision. The main task of the local controls is the responsibility of collection, while the central control is responsible for the management and guidance of all departments and units, as well as the collection of information from various departments. This information is located in the central control room and, in some cases, orders are sent to the parts [76].

The four levels are:

- MCC: from the MCC panels located in the panel room;
- Local: from the local control panels of each unit;
- Central: from the control room; and
- Automatic.

Each of the buildings and treatment units, individually or sometimes as a group, has a local equipment system. Input/output card control equipment and communication modules, as well as protective devices and local indicators, will be installed inside this panel. This panel is installed inside or near electrical panels.

All electric motors have a box installed next to the equipment, on which there is a selector switch, Start, Stop PBs, and lights indicating the status of these motors, including on, off, and failure. These indicators are installed in cases where the signs are installed. If the place is close to the equipment, the box can be installed on the local board.

The signaling related to the electric actuators of the valves is as follows:

- Open;
- Close;
- Fault (torque);
- Fault (temperature);
- Local/off/remote;
- Mcc/CR;
- Command open;
- Command close.

The signaling related to each engine or pump is as follows:

Run/Stop, Fault, Local/Off/Remote, Mcc/CR, Start/Stop command

- The status of the output juices should be entered into the control system and considered as digital inputs.
- Local control system panels are equipped with electric panels that have input and output cards and necessary communication connections. Their scope of action is determined below.
- All the precision tools used in the design and mentioned in the control philosophy are required to read the values in analog form, while those that act as switches are considered digital.
- All the analog values should be visible inside the system, in addition to being visible on the local control panel.
- The analog and digital information that enters the local PLC is transferred to the center through the communication BUS.
- The set of panels, if they are installed outside the electrical panel, depends on the type and size of the equipment in which they are installed. Factory and local tests of the panels should also be carried out according to the standards of their production. All related work such as installation, commissioning, alignment, wiring, numbering, and naming should be done. In addition, the specifications of construction works such as foundation, foundations, channels, etc., should be executed in time in connection with the installation of panels and control equipment and precision instruments. All the equipment and panels must be selected primarily and then selected as a group, and they should be connected to each other, tested, and started.
- Interposing relays should also be considered for the outputs that enter the electrical control system.
- The information from different units is collected in the control board that is placed in the control room.
- The control system includes all parts of the refinery.
- Values of all level gauges, performing meters such as pressure gauges and flow gauges, as well as level gauge switches.

The difference in level, switches for measuring the levels of sludge, and all precision instruments must be available on the local and central control panels.

1.1.33 SCADA SYSTEM DESIGN

The purpose of this system is to enable control and monitoring of the treatment plan from the monitoring center. This system must be open, meaning that one or more control systems may be added in the future. This system should also be open for communication with programming languages. The system in operation is connected to the monitoring.

Engineers can monitor the process. PLC, SCADA electronic systems must be grounded.

According to the definition, any obstacle placed in the path of the channel flow causes the water to back up and accelerate. If it is added while flowing over it, it is called a weir. Weirs can be studied in two states: free and submerged. Weirs can have sharp-edged or broad-edged profiles. Series weirs with rectangular, triangular, and trapezoidal cross-sections are constructed in all or part of the flow's cross-sectional area.

Similarly, a set of issues such as water scarcity, increasing population growth, proper sewage disposal, and surface runoff in cities are among the serious challenges related to water scarcity.

Due to the limited water resources, it is important to preserve the existing water sources and prevent them from being polluted. Sewage generated in residential and industrial sectors is one of the main sources of water pollution if not properly managed. Due to the population dispersion in cities, it becomes even more challenging to prevent the pollution of existing water sources.

One of the important issues alongside water scarcity and demand is the collection and treatment of wastewater for the purpose of public health, environmental order, and wastewater reuse. In this regard, several advancements have been made in the field of intelligent water purification, storage, transmission, and distribution facilities, as well as wastewater collection, treatment, and disposal facilities.

With the expansion of residential communities and the increase in population on one hand, and the expansion of industries and factories on the other hand, the issue of environmental pollution is becoming more important day by day. It makes the lives of humans and animals more unhealthy, putting them in more danger. Therefore, it is necessary to improve the collection of sewage and wastewater from the environment in residential communities.

Wastewaters always contain a large amount of foreign substances that are harmful to living organisms in various ways. The presence of pathogenic bacteria and germs in sewage is a factor that has become obvious to people over the past 100 years. The deaths caused by infectious diseases such as cholera, plague, dysentery, etc., in various parts of the world in the past, have put people at risk due to the contamination of water with these microbes. If untreated sewage enters the environment and natural water sources, it causes these sources to be contaminated with pathogenic microbes. Human contact with these sources increases the risk of spreading various diseases among people.

Disruption of environmental order is mostly caused by the spreading of sewage and surface water on public roads. Rainwater, after washing the streets and being contaminated with organic and mineral substances on their surface in the form of surface sewage, covers the streets and causes disruption of order, creating an undesirable landscape in the human living environment. The entry of such sewage into the environment, in addition to the direct dangers it poses to people's health and creating ugly landscapes, can also lead to unpleasant odors and an increase in annoying insects, especially mosquitoes and flies, which also threaten human health.

Collecting wastewater has a significant effect on lowering the level of underground water and maintaining the natural resources of underground water. The issue of not polluting the underground water becomes more important in some parts of the world where the sewage disposal system is carried out in its traditional way using absorbent wells. In these places, due to the lack of access to natural sources of surface water such as rivers or freshwater lakes, it is common to use underground water for water supply. Therefore, in addition to being the main cause of groundwater pollution, the use of sewage-absorbing wells also affects the groundwater level and causes soil pollution, creating problems in terms of construction in low-lying areas and pits.

The discharge of untreated sewage into running water causes water sources to be depleted of dissolved oxygen and makes life difficult for aquatic life, so fish and other aquatic animals will not be able to live in such waters. Fish need at least 3 to 5 milligrams per liter of dissolved oxygen to survive. The discharge of wastewater containing suspended and organic substances into surface waters increases the turbidity and reduces the passage of sunlight to the depths of the water, which is effective in reducing the growth of plants, especially algae. Algae are a source of dissolved oxygen production in water sources. The sedimentation of suspended materials on the riverbed has gradually led to the formation of a layer of sludge, making the place unsuitable for fish spawning. In addition, anaerobic biological activities occur in the sludge layer in the riverbed due to the lack of oxygen, resulting in the production of unpleasant gases and the discoloration of the water. Such water cannot be used for various purposes without purification.

1.1.34 SMARTENING THE PROCESS OF OBTAINING WATER FROM SEWAGE

Due to the increasing need for water, the issue of the reuse of wastewater is receiving more attention day by day. Especially in dry countries due to water scarcity and high cost of drinking water, the use of treated wastewater for non-domestic uses such as irrigation of green spaces, parks, afforestation, and washing of streets and sewer channels becomes particularly important. Here, it should be noted that the use of wastewater refined for drinking purposes, due to economic reasons and lack of acceptance by the general public, has not yet taken a wide and practical aspect in the world. If the reuse of purified wastewater and sending it in the urban network for irrigation of green spaces and industrial uses, and is used day by day in more parts of the world. Because the amount of dissolved mineral salts in wastewater is far less than in open seawater and wastewater produced from freshwater for agricultural irrigation is far cheaper than the desalination of salt lake water, this is a problem in dry countries where people in many places are facing a shortage of freshwater, it can save the consumption of freshwater used in agricultural irrigation. It should be noted here that using raw and untreated wastewater for agricultural irrigation causes many problems and often makes the use of such wastewater impossible. Today, smart technologies for water and wastewater systems such as networked sensors, Geospatial Information Systems (GIS), Remote Sensing (RS), and the Internet of Things (IoT) play an important role in the reuse of wastewater.

1.1.35 NETWORKED SENSORS, GIS, RS, AND IOT FOR REUSE OF WASTEWATER

The present work deals with the control of environmental contaminants and the reuse of wastewater. Smart technologies including networked sensors, GIS, RS, and IoT provide modern techniques for the reuse of wastewater.

1.2 METHODS

In this work, the research method was stated in the form of the following chart (Figure 1.1).

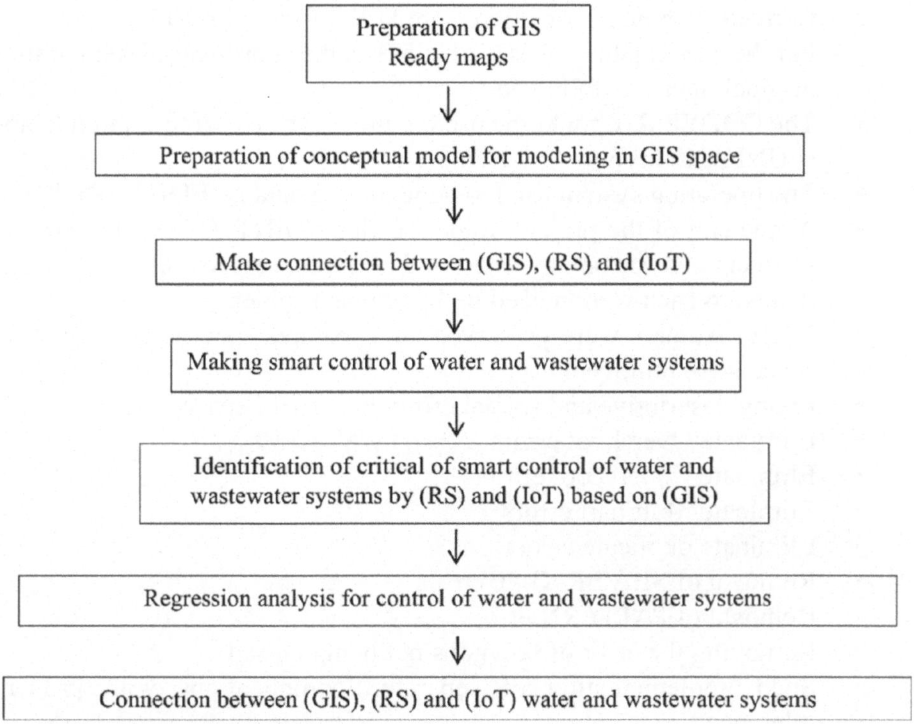

FIGURE 1.1 Research flow chart.

In this work, the following hypotheses can be expected due to intercommunication techniques for water and wastewater:

- Does the use of intercommunication techniques increase the smart control of water and wastewater systems?
- Is it possible to identify the critical zone of water and wastewater systems by GIS, RS, and IoT?
- Why can the rapid data intercommunication technique be able to find the critical zone?
- How can conceptual modeling predict the failure of water and wastewater systems?

For providing a GIS platform and georeferenced data, firstly, the preparation of GIS Ready maps of the facilities in the water supply model was carried out as follows:

- Exchange graphic information from CAD space to GIS space
- Fixed errors in the CAD space

- Convert graphic information from DWG format to SHP
- For the coordinate system of shapefiles, the coordinate system of the original map was followed.
- The CODEPAGE font type used in the descriptive information table is UNICODE type.
- The operating system used to generate the data is WINDOWS XP.
- At the end of the project, folders called POINT, LINE, and AREA were created for points, lines, and surfaces, respectively, and maps related to each were placed in the relevant folder.
- Useful features were preserved, and features specific to the CAD space were removed.
- Fixing descriptive and spatial errors in the GIS space.
- Eliminate Overshoot errors
- Eliminate Undershoot errors
- Eliminate Multipart errors
- Eliminate duplicate errors
- Removal of SILVER–GAP errors
- Removal of PATTERN errors
- Removing the error of polygons not being closed
- Snap implementation of separated complications with proper tolerance.
- Removing toll errors that were in the wrong place.
- Create primary and foreign keys for the tax table.
- Establishing proper tolerance and exchange of complications from the topology space.
- Preparing a conceptual model for transmission line modeling in GIS space.
- The ERD (Entity-Relationship Diagram) model is desired, and in this model, information is expressed by the three basic concepts of entity, attribute, and relationship (including the type and kind of relationship).
- Complete the characteristics of each complication.
- Preparation of a suitable name for each field.
- Completing the fee tables based on the prepared forms.
- Create a proper land database.

The proposed geodatabase includes a database that is supported by ArcGIS and SQL Server and is programmable. The process of this work includes the following items due to database features:

- Creating domains for descriptions (attribute domains) necessary to avoid entering false descriptions into the database (by assigning description codes).
- Creating different categories (datasets) for complications for different management purposes.
- Automatic updating of sizes and areas after each change (dimension update).
- Creating a spatial reference for complications and also creating spatial domains for complications in order to prevent the entry of extraneous spatial data.
- Creation of relationship classes between spatial and non-spatial data for the spatial relationship of complications, so that each complication is related according to the type of relationship (simple or compound) and its degree (cardinality). With these classes, the leakage points of the transmission line can be identified.
- Creating subgroups for complications to create validation rules (Validate rules) and also prevent additional queries in the database.
- Creating an index for descriptive tables to increase the accuracy and speed of queries.
- Creating topology rules to prevent wrong edits, such as separation of equipment from pipes, such as separation of valve from pipe or manhole from pipe.
- Creation of Annotation classes for immediate calling of descriptions in order to print them.
- Determining the equipment needed to solve the accident, by having available the type of valve or pipe involved in the accident.
- Creating a smart network of water transmission lines by introducing charges in the network (Networking)

Creating the ability to track and implement network analysis, including the following:

- Descriptive information about the location of the valves that must be closed to cut off the water in the accident area.
- Descriptive information about the location of pipes that become dehydrated due to closing a transmission line valve.
- Descriptive location information of common valves of pipes.
- Descriptive information about the location of the pipes downstream of the manhole accident.
- Descriptive information about the location of the pipes above the manhole in the accident (caught pipes).

- Increasing the speed of dealing with incidents and reducing physical losses of water.
- Systematizing the storage and use of distribution network equipment information using GIS.
- Economic savings due to reducing follow-up of events.
- Creating maps and reporting.
- Combination of all types of analyses for optimal and immediate management of facilities.

In this work, after preparing GIS-ready, water supply model facility maps were selected in order to achieve the following goals:

- Estimation and evaluation of water distribution system.
- Water quality studies.
- Studies to prevent wastage of water and energy.
- Improvement and development of water distribution system.
- Connecting with nearby networks.
- Update existing facilities.

The process of choosing the model and the type of software to prepare the hydraulic model is as follows:

- Hydraulic analysis;
- Water quality analysis;
- Simulation features and specifications;
- Graphic features;
- Hardware requirements;
- Software requirements;
- Software capabilities;
- Required data;
- Cost.

The information needed in the preparation of the hydraulic model of the water distribution system includes information about pipes, faucets, pumps, nodes, operating conditions, water consumption and needs, consumption fluctuations and water consumption patterns in the city, and loss coefficient. Therefore, quantitative information required for hydraulic modeling and analysis was prepared and studied under three branches as described below:

- Network information;
- Information related to exploitation;
- Information about consumption.

Pipes information of network are as the following items:

- Pipe code;
- Flow direction in the pipe;
- Node code of the beginning and end of the pipe;
- Diameter;
- Material;
- Length;
- Roughness coefficient;
- Coefficient C;
- Partial loss coefficients;
- Working pressure and design pressure.

The information about the pumps are as follows:

- Start and end node code;
- The diameter of the vane suction pipe;
- Pump head;
- Pump capacity;
- Altitude number of the place;
- Pump installation height figure.

1.3 RESULTS AND DISCUSSION

In the present work, the hydraulic model was defined in the form of a water distribution system analysis in accordance with GIS and Water Gems, in compliance with ArcGIS-ArcMap (Tables 1.2–1.5). Graphs of the pressure fluctuations measured by remote sensing in accordance with Water Gems and ArcGIS-ArcMap are shown in (Figures 1.2–1.4).

1.4 CONCLUSION

In this work, the aim was to facilitate quick data intercommunication for water and wastewater facilities. Various complications, classes, and sub-classes related to supply, collection, storage, transmission, and distribution facilities were investigated in order to create a GIS-based location platform. This allowed for better management and control of urban facilities using GIS and fuzzy logic.

The results of this work emphasized the following points:

- Management of water consumption and demand.

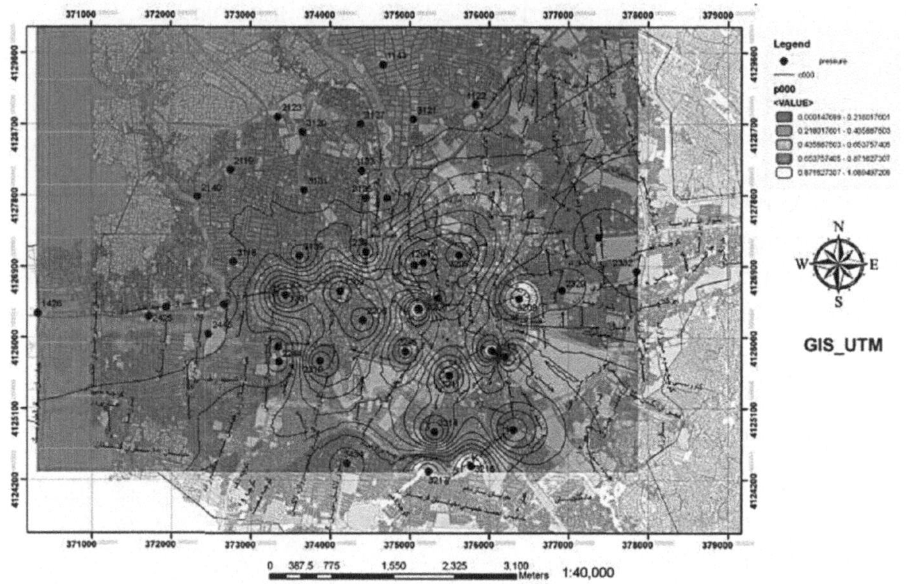

FIGURE 1.2 Graphs of the pressure fluctuations measurement by remote sensing in accordance with WaterGems and ArcGIS-ArcMap.

TABLE 1.2 Water Distribution Network Model

Start Node	Stop Node	Material	Hazen–Williams (C)	Flow (lit/s)	Velocity (m/s)
J-1	J-2	PE	130/0	0.945	0.33
J-3	J-4	PE	130/0	0.063	0.04
J-5	J-6	AC	130/0	8.883	1.12
J-7	J-8	AC	130/0	0.315	0.006
J-9	J-10	AC	130/0	–4.219	0.005
J-11	J-12	AC	130/0	0.252	0.03
J-13	J-14	AC	130/0	–4.158	0.533
J-15	J-16	PE	130/0	1.638	0.119
J-17	J-18	PE	130/0	–0.63	0.09
J-19	J-20	AC	130/0	0.252	0.033
J-21	J-22	PE	130/0	2.394	0.37
J-23	J-24	AC	130/0	1.386	0.17
J-25	J-26	AC	130/0	0.189	0.16
J-27	J-28	AC	130/0	2.394	0.304
J-29	J-30	PE	130/0	–1.89	0.302
J-31	J-32	AC	130/0	0.252	0.037
J-33	J-34	AC	130/0	21.42	0.43

TABLE 1.2 (Continued)

Start Node	Stop Node	Material	Hazen–Williams (C)	Flow (lit/s)	Velocity (m/s)
J-35	J-36	AC	130/0	–4.158	0.52
J-37	J-38	PE	130/0	3.528	0.55
J-39	J-40	AC	130/0	–4.473	0.57
J-41	J-42	PE	130/0	1.008	0.15
J-43	J-44	PE	130/0	0.126	0.015
J-45	J-46	AC	130/0	–0.945	0.122
J-47	J-48	AC	130/0	1.575	0.19
J-49	J-50	AC	130/0	–6.363	0.131
J-51	J-14	AC	130/0	–12.41	1.57
J-52	J-53	AC	130/0	0.441	0.056
J-54	J-55	AC	130/0	2.457	0.304
J-56	J-57	AC	130/0	0.882	0.116
J-58	J-59	AC	130/0	–1.071	0.137
J-60	J-61	PE	130/0	0.63	0.104
J-62	J-63	PE	130/0	–4.284	0.677
J-64	J-65	PE	130/0	1.575	0.53
J-66	J-67	AC	130/0	0.126	0.012
J-68	J-69	PE	130/0	–0.252	0.039
J-70	J-71	PE	130/0	0.126	0.037
J-72	J-73	AC	130/0	–0.945	0.119
J-74	J-15	PE	130/0	–0.126	0.012
J-75	J-76	PE	130/0	0.252	0.076
J-77	J-78	AC	130/0	–4.725	0.597
J-81	J-82	PE	130/0	–0.693	0.107
J-83	J-84	AC	130/0	0.126	0.015
J-85	J-86	AC	130/0	20.727	0.66
J-87	J-88	AC	130/0	–2.52	0.317
J-89	J-90	AC	130/0	0.378	0.046
J-91	J-92	AC	130/0	2.016	0.256
J-93	J-94	AC	130/0	0.819	0.104
J-95	J-96	AC	130/0	1.701	0.219
J-97	J-98	AC	130/0	–1.323	0.167
J-59	J-99	AC	130/0	0.063	0.006
J-100	J-101	AC	130/0	1.071	0.134

TABLE 1.3 Water Distribution Network Model

ID	Label	Pressure (m-h2o)
31	J-1	36.5
32	J-2	37.2
34	J-3	35.9
35	J-4	36.5
37	J-5	37.2
38	J-6	38.6
40	J-7	39.3
41	J-8	40
43	J-9	40.7
44	J-10	40.7
46	J-11	38.6
47	J-12	37.9
49	J-13	40.7
50	J-14	39.3
52	J-15	37.2
53	J-16	35.9
55	J-17	36.5
56	J-18	36.5
58	J-19	37.9
59	J-20	38.6
61	J-21	37.2
62	J-22	36.5
64	J-23	37.9
65	J-24	38.6
67	J-25	40.7
68	J-26	42.1
70	J-27	40
71	J-28	39.3
73	J-29	38.6
74	J-30	37.2
76	J-31	36.5
77	J-32	37.2
79	J-33	40.7
80	J-34	41.4
82	J-35	36.5
83	J-36	36.5

TABLE 1.3 *(Continued)*

ID	Label	Pressure (m-h2o)
85	J-37	36.5
86	J-38	36.5
88	J-39	38.6
89	J-40	38.6
91	J-41	36.5
92	J-42	35.9
94	J-43	36.5
95	J-44	36.5
97	J-45	41.4
98	J-46	41.4
100	J-47	36.5

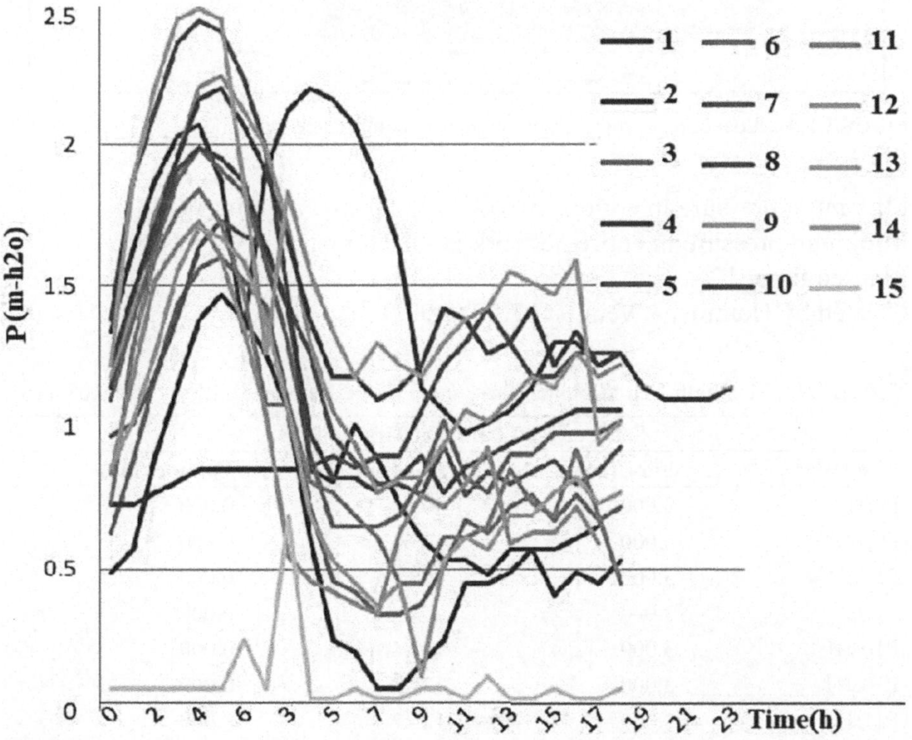

FIGURE 1.3 Graphs of the pressure fluctuations measured by remote sensing in accordance with water gems and ArcGIS-ArcMap.

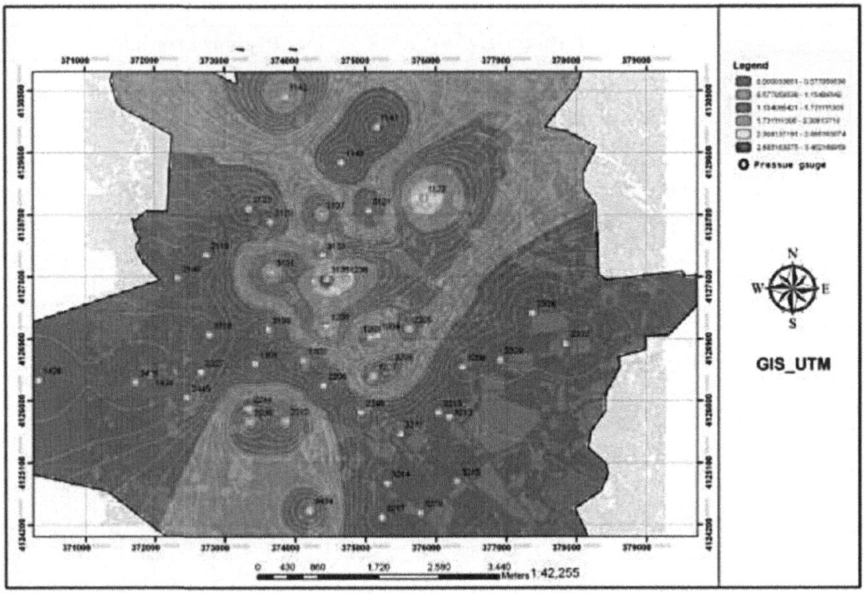

FIGURE 1.4 Locations of the pressure gauges in compliance with GIS.

Maximum pressure in entire network is 121.1 mH at point P6:43%
Minimum pressure in entire network is – 10.0 mH at point P24:80%
Elapsed time: 12 s.
Created by Hammer – Version 07.00.049.00

TABLE 1.4 Maximum and Minimum Pressure in Entire Network; 0.0148(S) to 0.0886(S)

Selected End Points at Time 0.0148 s			
End Point	**Flow (cms)**	**Head (m)**	**Volume (m3)**
P0:J1	3.000	40.6	0.000
P0:J2	3.000	40.6	0.000
P1:J2	3.000	40.6	0.000
P1:J3	3.000	40.6	0.000
P10:J10	3.000	127.4	0.000
P10:J11	3.000	125.8	0.000
P11:J11	3.000	125.7	0.000
P11:J12	3.000	123.0	0.000
P12:J12	3.000	123.0	0.000
P12:J13	3.000	120.4	0.000

TABLE 1.4 *(Continued)*

Selected End Points at Time 0.0148 s			
End Point	**Flow (cms)**	**Head (m)**	**Volume (m3)**
P13:J13	3.000	120.3	0.000
P13:J14	3.000	120.0	0.000
P14:J14	3.000	119.9	0.000
P14:J15	3.000	115.2	0.000
P15:J15	3.000	115.2	0.000
P15:J16	3.000	114.9	0.000
P16:J16	3.000	114.8	0.000
P16:J17	3.000	113.0	0.000
P17:J17	3.000	113.0	0.000
P17:J18	3.000	112.9	0.000
P18:J18	3.000	112.8	0.000
P18:J19	3.000	112.8	0.000
P19:J19	3.000	112.7	0.000
P19:J20	3.000	112.7	0.000
P2:J6	3.000	133.3	0.000
P2:J7	3.000	132.8	0.000
P20:J20	3.000	112.6	0.000
P20:J21	3.000	108.6	0.000
P21:J22	3.000	106.7	0.000
P22:J22	3.000	106.6	0.000
P22:J23	3.000	105.4	0.000
P23:J23	3.000	105.4	0.000
P23:J24	3.000	105.2	0.000
P24:J24	3.000	105.2	0.000
P24:J28	3.000	104.6	0.000
P25:J28	3.000	104.5	0.000
P25:N1	3.000	104.3	0.000
P3:J7	2.500	132.8	0.000
P3:J8	2.500	130.3	0.000
P4:J3	3.000	135.0	0.000
P4:J4	3.000	135.0	0.000
P5:J26	3.000	135.0	0.000
P5:J4	3.000	135.0	0.000
P6:J26	3.000	134.9	0.000
P6:J27	3.000	133.5	0.000

TABLE 1.4 *(Continued)*

Selected End Points at Time 0.0148 s			
End Point	**Flow (cms)**	**Head (m)**	**Volume (m3)**
P7:J27	3.000	133.5	0.000
P7:J6	3.000	133.3	0.000
P8:J8	3.000	130.3	0.000
P8:J9	3.000	130.2	0.000
P9:J10	3.000	127.5	0.000
P9:J9	3.000	130.1	0.000

TABLE 1.5 Maximum and Minimum Pressure in Entire Network; 0.0148(S) to 0.0886(S)

Selected End Points at Time 0.0295 s			
End Point	**Flow (cms)**	**Head (m)**	**Volume (m3)**
P0:J1	3.000	40.6	0.000
P0:J2	3.000	40.6	0.000
P1:J2	3.000	40.6	0.000
P1:J3	3.000	40.6	0.000
P10:J10	3.000	127.4	0.000
P10:J11	3.000	125.8	0.000
P11:J11	3.000	125.7	0.000
P11:J12	3.000	123.0	0.000
P12:J12	3.000	123.0	0.000
P12:J13	3.000	120.4	0.000
P13:J13	3.000	120.3	0.000
P13:J14	3.000	120.0	0.000
P14:J14	3.000	119.9	0.000
P14:J15	3.000	115.2	0.000
P15:J15	3.000	115.2	0.000
P15:J16	3.000	114.9	0.000
P16:J16	3.000	114.8	0.000
P16:J17	3.000	113.0	0.000
P17:J17	3.000	113.0	0.000
P17:J18	3.000	112.9	0.000
P18:J18	3.000	112.8	0.000
P18:J19	3.000	112.8	0.000
P19:J19	3.000	112.7	0.000
P19:J20	3.000	112.7	0.000
P2:J6	3.000	133.3	0.000

TABLE 1.5 *(Continued)*

Selected End Points at Time 0.0295 s			
End Point	**Flow (cms)**	**Head (m)**	**Volume (m3)**
P2:J7	2.747	156.3	0.000
P20:J20	3.000	112.6	0.000
P21:J21	3.000	108.6	0.000
P21:J22	3.000	106.7	0.000
P22:J22	3.000	106.6	0.000
P22:J23	3.000	105.4	0.000
P23:J23	3.000	105.4	0.000
P23:J24	3.000	105.2	0.000
P24:J24	3.000	105.2	0.000
P24:J28	3.000	104.6	0.000
P25:J28	3.000	104.5	0.000
P25:N1	3.000	104.3	0.000
P3:J7	2.747	156.2	0.000
P3:J8	2.745	107.0	0.000
P4:J3	3.000	135.0	0.000
P4:J4	3.000	135.0	0.000
P5:J26	3.000	135.0	0.000
P5:J4	3.000	135.0	0.000
P6:J26	3.000	134.9	0.000
P6:J27	3.000	133.5	0.000
P7:J27	3.000	133.5	0.000
P7:J6	3.000	133.3	0.000
P8:J8	2.745	107.0	0.000
P8:J9	2.750	106.5	0.000
P9:J10	3.000	127.5	0.000
P9:J9	2.750	106.5	0.000
Selected End Points at Time 0.0443 s			
P0:J1	3.000	40.6	0.000
P0:J2	3.000	40.6	0.000
P1:J2	3.000	40.6	0.000
P1:J3	3.000	40.6	0.000
P10:J10	3.000	127.4	0.000
P10:J11	3.000	125.8	0.000
P11:J11	3.000	125.7	0.000
P11:J12	3.000	123.0	0.000

TABLE 1.5 *(Continued)*

	Selected End Points at Time 0.0295 s		
End Point	**Flow (cms)**	**Head (m)**	**Volume (m3)**
P12:J12	3.000	123.0	0.000
P12:J13	3.000	120.4	0.000
P13:J13	3.000	120.3	0.000
P13:J14	3.000	120.0	0.000
P14:J14	3.000	119.9	0.000
P14:J15	3.000	115.2	0.000
P15:J15	3.000	115.2	0.000
P15:J16	3.000	114.9	0.000
P16:J16	3.000	114.8	0.000
P16:J17	3.000	113.0	0.000
P17:J17	3.000	113.0	0.000
P17:J18	3.000	112.9	0.000
P18:J18	3.000	112.8	0.000
P18:J19	3.000	112.8	0.000
P19:J19	3.000	112.7	0.000
P19:J20	3.000	112.7	0.000
P2:J6	3.000	133.3	0.000
P2:J7	2.747	156.3	0.000
P20:J20	3.000	112.6	0.000
P20:J21	3.000	108.6	0.000
P21:J21	3.000	108.6	0.000
P21:J22	3.000	106.7	0.000
P22:J22	3.000	106.6	0.000
P22:J23	3.000	105.4	0.000
P23:J23	3.000	105.4	0.000
P23:J24	3.000	105.2	0.000
P24:J24	3.000	105.2	0.000
P24:J28	3.000	104.6	0.000
P25:J28	3.000	104.5	0.000
P25:N1	3.000	104.3	0.000
P3:J7	2.747	156.3	0.000
P3:J8	2.749	106.6	0.000
P4:J3	3.000	135.0	0.000
P4:J4	3.000	135.0	0.000
P5:J26	3.000	135.0	0.000

TABLE 1.5 *(Continued)*

	Selected End Points at Time 0.0295 s		
End Point	**Flow (cms)**	**Head (m)**	**Volume (m3)**
P5:J4	3.000	135.0	0.000
P6:J26	3.000	134.9	0.000
P6:J27	3.000	133.5	0.000
P7:J27	3.000	133.5	0.000
P7:J6	3.000	133.3	0.000
P8:J8	2.749	106.6	0.000
P8:J9	2.750	106.5	0.000
P9:J10	3.000	127.5	0.000
P9:J9	2.750	106.5	0.000
	Selected End Points at Time 0.0590 s		
P0:J1	3.000	40.6	0.000
P0:J2	3.000	40.6	0.000
P1:J2	3.000	40.6	0.000
P1:J3	3.000	40.6	0.000
P10:J10	3.000	127.4	0.000
P10:J11	3.000	125.8	0.000
P11:J11	3.000	125.7	0.000
P11:J12	3.000	123.0	0.000
P12:J12	3.000	123.0	0.000
P12:J13	3.000	120.4	0.000
P13:J13	3.000	120.3	0.000
P13:J14	3.000	120.0	0.000
P14:J14	3.000	119.9	0.000
P14:J15	3.000	115.2	0.000
P15:J15	3.000	115.2	0.000
P15:J16	3.000	114.9	0.000
P16:J16	3.000	114.8	0.000
P16:J17	3.000	113.0	0.000
P17:J17	3.000	113.0	0.000
P17:J18	3.000	112.9	0.000
P18:J18	3.000	112.8	0.000
P18:J19	3.000	112.8	0.000
P19:J19	3.000	112.7	0.000
P19:J20	3.000	112.7	0.000
P2:J6	3.000	133.3	0.000

TABLE 1.5 *(Continued)*

Selected End Points at Time 0.0295 s			
End Point	**Flow (cms)**	**Head (m)**	**Volume (m3)**
P2:J7	2.747	156.3	0.000
P20:J20	3.000	112.6	0.000
P21:J21	3.000	108.6	0.000
P21:J22	3.000	106.7	0.000
P22:J22	3.000	106.6	0.000
P22:J23	3.000	105.4	0.000
P23:J23	3.000	105.4	0.000
P23:J24	3.000	105.2	0.000
P24:J24	3.000	105.2	0.000
P24:J28	3.000	104.6	0.000
P25:J28	3.000	104.5	0.000
P25:N1	3.000	104.3	0.000
P3:J7	2.747	156.3	0.000
P3:J8	2.749	106.6	0.000
P4:J3	3.000	135.0	0.000
P4:J4	3.000	135.0	0.000
P5:J26	3.000	135.0	0.000
P5:J4	3.000	135.0	0.000
P6:J26	3.000	134.9	0.000
P6:J27	3.000	133.5	0.000
P7:J27	3.000	133.5	0.000
P7:J6	3.000	133.3	0.000
P8:J8	2.749	106.6	0.000
P8:J9	2.750	106.5	0.000
P9:J10	3.000	127.5	0.000
P9:J9	2.750	106.4	0.000
Selected End Points at Time 0.0738 s			
P0:J1	3.000	40.6	0.000
P0:J2	3.000	40.6	0.000
P1:J2	3.000	40.6	0.000
P1:J3	3.000	40.6	0.000
P10:J10	3.000	127.4	0.000
P10:J11	3.000	125.8	0.000
P11:J11	3.000	125.7	0.000
P11:J12	3.000	123.0	0.000

TABLE 1.5 *(Continued)*

Selected End Points at Time 0.0295 s			
End Point	**Flow (cms)**	**Head (m)**	**Volume (m3)**
P12:J12	3.000	123.0	0.000
P12:J13	3.000	120.4	0.000
P13:J13	3.000	120.3	0.000
P13:J14	3.000	120.0	0.000
P14:J14	3.000	119.9	0.000
P14:J15	3.000	115.2	0.000
P15:J15	3.000	115.2	0.000
P15:J16	3.000	114.9	0.000
P16:J16	3.000	114.8	0.000
P16:J17	3.000	113.0	0.000
P17:J17	3.000	113.0	0.000
P17:J18	3.000	112.9	0.000
P18:J18	3.000	112.8	0.000
P18:J19	3.000	112.8	0.000
P19:J19	3.000	112.7	0.000
P19:J20	3.000	112.7	0.000
P2:J6	2.704	152.7	0.000
P2:J7	2.747	156.3	0.000
P20:J20	3.000	112.6	0.000
P20:J21	3.000	108.6	0.000
P21:J21	3.000	108.6	0.000
P21:J22	3.000	106.7	0.000
P22:J22	3.000	106.6	0.000
P22:J23	3.000	105.4	0.000
P23:J23	3.000	105.4	0.000
P23:J24	3.000	105.2	0.000
P24:J24	3.000	105.2	0.000
P24:J28	3.000	104.6	0.000
P25:J28	3.000	104.5	0.000
P25:N1	3.000	104.3	0.000
P3:J7	2.747	156.3	0.000
P3:J8	2.749	106.6	0.000
P4:J3	3.000	135.0	0.000
P4:J4	3.000	135.0	0.000
P5:J26	3.000	135.0	0.000

TABLE 1.5 *(Continued)*

	Selected End Points at Time 0.0295 s		
End Point	**Flow (cms)**	**Head (m)**	**Volume (m3)**
P5:J4	3.000	135.0	0.000
P6:J26	3.000	134.9	0.000
P6:J27	3.000	133.5	0.000
P7:J27	3.000	133.5	0.000
P7:J6	2.704	152.7	0.000
P8:J8	2.749	106.5	0.000
P8:J9	2.750	106.5	0.000
P9:J10	3.000	127.5	0.000
P9:J9	2.750	106.4	0.000
	Selected End Points at Time 0.0886 s		
P0:J1	3.000	40.6	0.000
P0:J2	3.000	40.6	0.000
P1:J2	3.000	40.6	0.000
P1:J3	3.000	40.6	0.000
P10:J10	3.000	127.4	0.000
P10:J11	3.000	125.8	0.000
P11:J11	3.000	125.7	0.000
P11:J12	3.000	123.0	0.000
P12:J12	3.000	123.0	0.000
P12:J13	3.000	120.4	0.000
P13:J13	3.000	120.3	0.000
P13:J14	3.000	120.0	0.000
P14:J14	3.000	119.9	0.000
P14:J15	3.000	115.2	0.000
P15:J15	3.000	115.2	0.000
P15:J16	3.000	114.9	0.000
P16:J16	3.000	114.8	0.000
P16:J17	3.000	113.0	0.000
P17:J17	3.000	113.0	0.000
P17:J18	3.000	112.9	0.000
P18:J18	3.000	112.8	0.000
P18:J19	3.000	112.8	0.000
P19:J19	3.000	112.7	0.000
P19:J20	3.000	112.7	0.000
P2:J6	2.704	152.7	0.000

TABLE 1.5 *(Continued)*

Selected End Points at Time 0.0295 s			
End Point	**Flow (cms)**	**Head (m)**	**Volume (m3)**
P2:J7	2.747	156.3	0.000
P20:J20	3.000	112.6	0.000
P20:J21	3.000	108.6	0.000
P21:J21	3.000	108.6	0.000
P21:J22	3.000	106.7	0.000
P22:J22	3.000	106.6	0.000
P22:J23	3.000	105.4	0.000
P23:J23	3.000	105.4	0.000
P23:J24	3.000	105.2	0.000
P24:J24	3.000	105.2	0.000
P24:J28	3.000	104.6	0.000
P25:J28	3.000	104.5	0.000
P25:N1	3.000	104.3	0.000
P3:J7	2.747	156.3	0.000
P3:J8	2.749	106.6	0.000
P4:J3	3.000	135.0	0.000
P4:J4	3.000	135.0	0.000
P5:J26	3.000	135.0	0.000
P5:J4	3.000	135.0	0.000
P6:J26	3.000	134.9	0.000
P6:J27	3.000	133.5	0.000
P7:J27	3.000	133.5	0.000
P7:J6	2.704	152.7	0.000
P8:J8	2.749	106.5	0.000
P8:J9	2.750	106.5	0.000
P9:J10	3.000	127.5	0.000
P9:J9	2.750	106.4	0.000

- dentification of unauthorized branches with high water consumption.
- Determining the location of water sources for green spaces and separating irrigation systems for green spaces from drinking water.
- Separating irrigation systems for green spaces from drinking water.
- Scheduling irrigation during non-peak hours.
- Identifying subscribers with an average consumption higher than half a liter per second and negotiating with them to shift their consumption to non-peak hours.

- Utilizing reservoirs as a tool to control and manage network peak.
- By reducing the network pressure, the stability of the network should be maintained.
- Network pressure management through the use of tanks, pressure relief valves, flow control valves, and the installation of pressure gauges and zoning loggers.
- Installation of tools to improve consumption.
- Installing a meter at the entrance of the water and wastewater network to detect leaks, which can then be repaired and necessary measures taken to reduce them.
- Installing a timer in the control circuit of the electric pump for wells and pumping stations to prevent tank overflow in appropriate locations.
- Using smart pilots on pressure relief valves to manage downstream network pressure during periods of minimum consumption.

ACKNOWLEDGMENT

The author thanks all specialists for their valuable observations and advice, as well as the referees for their recommendations that improved the quality of this paper.

KEYWORDS

- **fatigue index**
- **fuzzy logic**
- **geospatial information system**
- **internet of things**
- **networked sensors**
- **optimal bitumen percentage**
- **reclaimed asphalt pavement**
- **remote sensing**

REFERENCES

1. Zargar, M., Ahmadinia, E., Asli, H., & Karim, M. R., (2012). Investigation of the possibility of using waste cooking oil as a rejuvenating agent for aged bitumen. *Journal of Hazardous Materials, 233*, 254–258.
2. Hariri, A. K., Soltan, A. O., Thomas, S., Deepu, A., & Hariri, A. H., (2017). *Handbook of Research for Fluid & Solid Mechanics.* Published by the Apple Academic Press, Inc., ISBN 9781771885010, USA, Canada. https://doi.org/10.1201/9781315365701; https://www.taylorfrancis.com/books/edit/10.1201/9781315365701/handbook-research-fluid-solid-mechanics-kaveh-hariri-asli-soltan-ali-ogli-aliyev-sabu-thomas-deepu-gopakumar (accessed on 24 April 2024).
3. William, H. K., & Judith, M. T., (1978). Linear hypotheses. *International Encyclopedia of Statistics* (Vol. 1). Free Press.
4. Evan, J. W. (1972). "*I. Regression* (pp. 523–541).
5. Julian, C. S. (2023). "*II. Analysis of Variance* (pp. 541–554).
6. Asli, H. K., Khodaparast, H. A., Asli, H. H., & Sabermaash, E. E., (2012). Water hammer modeling and simulation by GIS. *Modeling and Simulation in Engineering, 1*(3), 4. Hindawi Publishing Corporation, Article ID 704163. doi: 10.1155/2012/704163, ISSN: 16875605, 16875591, Pub., USA. http://www.hindawi.com/journals/mse/aip/704163/ (accessed on 24 April 2024).
7. Lindley, D. V., (1987). Regression and correlation analysis. *New Palgrave: A Dictionary of Economics, 4*, 120–123.
8. Birkes, D., & Dodge, Y. (1993). *Alternative Methods of Regression*. ISBN 0-471-56881-3.
9. Gardziejczyk, W., (2002). *Texture of Road Surfaces, Methods of Measurement Parameters Evaluation and its Influence on the Tire/Road Noise*. Drogi Mosty, no. 2, Poland.
10. Kian, H. A., (2024). Technological advancement in HVAC&R system energy efficiency. *Technological Advancement in Clean Energy Production.* Apple Academic Press, Hard ISBN: 9781774915585. https://www.appleacademicpress.com/title.php?id=1416 (accessed on 24 April 2024).
11. Kian, H. A., (2024). IoT-based smart water leak detection system for a sustainable future: A case study. *Sustainable Water Engineering.* Smart and Emerging Technologies. Apple Academic Press, Hard ISBN: 9781774915714. https://appleacademicpress.com/sustainable-water-engineering-smart-and-emerging-technologies/9781774915714 (accessed on 24 April 2024).
12. Amritanshu, S., Kian, H. A., Neha, K. R., Ann, R. A., & Haghi, A. K., (2024). *Technological Advancement in Clean Energy Production.* Apple Academic Press, Hard ISBN: 9781774915585. https://www.appleacademicpress.com/title.php?id=1416 (accessed on 24 April 2024).
13. Ali, P., Kian, H. A., Ann, R. A., & Haghi, A. K., (2024). *Sustainable Water Engineering; Smart and Emerging Technologies*. Apple Academic Press, Hard ISBN: 9781774915714. https://appleacademicpress.com/sustainable-water-engineering-smart-and-emerging-technologies/9781774915714 (accessed on 24 April 2024).
14. Kian, H. A., & Kaveh, H. A., (2022). Isolated pressure zones based on GIS as a solution for water network problems. *Water Practice and Technology*, wpt2022119. doi: https://doi.org/10.2166/wpt.2022.119, https://iwaponline.com/wpt/article/doi/10.2166/

wpt.2022.119/91338/Isolated-pressure-zones-based-on-GIS-as-a-solution (accessed on 24 April 2024).

15. Hossein, H. A. A. P., Ann, R. A., & Haghi, A. K., (2023). *New Advances in Materials Technologies; Experimental Characterizations, Theoretical Modeling, and Field Practices.* Hard ISBN: 9781774914847, https://www.appleacademicpress.com/new-advances-in-material-technologies-experimental-characterizations-theoretical-modeling-and-field-practices/9781774914847 (accessed on 24 April 2024).
16. Hossein, H. A., Tamara, T., Ann, R. A., & Haghi, A. K. (2024). *Sustainable Water Treatment and Ecosystem Protection Strategies.* Hard ISBN: 9781774915189, https://www.appleacademicpress.com/sustainable-water-treatment-and-ecosystem-protection-strategies-/9781774915189 (accessed on 24 April 2024).
17. Asli, H. H., Arabani, M., & Golpour, Y., (2020). Reclaimed asphalt pavement (RAP) based on a geospatial information system (GIS). *Slovak Journal of Civil Engineering, 28*(2), 36–42. Slovak University of Technology in Bratislava. Slovak. https://doi.org/10.2478/sjce-2020-0013.
18. Asli, H. H., & Arabani, M., (2022). Analysis of strain and failure of asphalt pavement. Computational research progress in applied science & engineering. *Transactions of Civil and Environmental Engineering, 8,* 1–11. Article ID: 2250. https://doi.org/10.52547/crpase.8.1.2250.
19. Asli, H. H., (2023). Reclaimed asphalt pavement (RAP) based on geospatial information system (GIS) and networked sensors modeling. *New Advances in Materials Technologies; Experimental Characterizations. Theoretical Modeling. and Field Practices*. Hard ISBN: 9781774914847. https://www.appleacademicpress.com/new-advances-in-material-technologies-experimental-characterizations-theoretical-modeling-and-field-practices/9781774914847 (accessed on 24 April 2024).
20. Asli, H. K., Nagiyev, F. B., Haghi, A. K., Aliyev, S. A., & Asli, H. H., (2010). Nonlinear heterogeneous model for water hammer disaster. *J. Balkan Tribological Association, 16*(2). 209–222.
21. Asli, K. H., Aliyev, S. A. O., Thomas, S., & Deepu, A., (2017). *Handbook of Research for Fluid and Solid Mechanics: Theory, Simulation, and Experiment.* Apple Academic Press. Taylor & Francis Group. New York. USA. https://doi.org/10.1201/9781315365701.
22. Nayar, V. (2013). The water crisis—Rethinking water governance. *Journal of Land and Rural Studies, 1*(1), 75–94.
23. Prakash, A., Singh, S., & Brouwer, L. (2015). Water transfer from peri-urban to urban areas: conflict over water for Hyderabad city in South India. *Environment and Urbanization ASIA, 6*(1), 41–58.
24. Hariri, A. H., & Hozori, A., (2021). Non-revenue water (NRW) and 3D hierarchical model for landslide. *Larhyss Journal, 48,* 189–210. http://larhyss.net/ojs/index.php/larhyss/article/view/810/810 (accessed on 24 April 2024).
25. Hariri, A. H., (2022). Investigation of the factors affecting pedestrian accidents in urban roundabouts. *Computational Research Progress in Applied Science & Engineering (CRPASE), 8*(1). 1–4. Article ID: 2255. https://crpase.com/archive/CRPASE-Vol-08-issue-01-15172135.pdf; https://doi.org/10.52547/crpase.8.1.2255 (accessed on 24 April 2024).
26. Hariri, A. H., & Nazari, S., (2021). Water age and leakage in reservoirs: Some computational aspects and practical hints. *Larhyss Journal, 48*, 151–167. http://larhyss.net/ojs/index.php/larhyss/article/view/808/807 (accessed on 24 April 2024).

27. Kian, H. A., & Kaveh, H. A., (2023). Minimum night flow (MNF) and corrosion control in compliance with internet of things (IoT) for water systems. *Water Practice and Technology*. wpt2023012, https://doi.org/10.2166/wpt.2023.012; https://iwaponline.com/wpt/article/doi/10.2166/wpt.2023.012/93513/Minimum-night-flow-MNF-and-corrosion-control-in (accessed on 24 April 2024).
28. Phipps, M., & Brace-Govan, J. (2011). From right to responsibility: Sustainable change in water consumption. *Journal of Public Policy & Marketing, 30*(2), 203–219.
29. Shahrokhi, M., Feizbahr, M., Choong, K. K., & Rostami, F., (2018). Wave energy dissipation using perforated and non-perforated piles. *International Journal of Engineering Transactions B: Applications, 31*(2), 212–219.
30. Feizbahr, M., Mirhosseini, S. M., & Joshaghani, A. H., (2020). Improving the performance of conventional concrete using multi-walled carbon nanotubes. *Express Nano Letters, 1*(1), 1–9.
31. Asli, H. H., (2023). Applications of networked sensors and the internet of things (IoT) for wastewater treatment. *Sustainable Water Treatment and Ecosystem Protection Strategies*. Hard ISBN: 9781774915189. https://www.appleacademicpress.com/sustainable-water-treatment-and-ecosystem-protection-strategies-/9781774915189 (accessed on 24 April 2024).
32. Asli, H. H., (2023). Applications of networked sensors and the internet of things (IoT) for water treatment. *Sustainable Water Treatment and Ecosystem Protection Strategies*. Hard ISBN: 9781774915189. https://www.appleacademicpress.com/sustainable-water-treatment-and-ecosystem-protection-strategies-/9781774915189 (accessed on 24 April 2024).
33. Asli, H. H., (2023). Modeling of corrosion for water system by networked sensors and the internet of things (IoT) in compliance with geography information system (GIS). *Sustainable Water Treatment and Ecosystem Protection Strategies*. Hard ISBN: 9781774915189. https://www.appleacademicpress.com/sustainable-water-treatment-and-ecosystem-protection-strategies-/9781774915189 (accessed on 24 April 2024).
34. Bergua Amores, J. A. (2008). Ideology, Magic and Spectres: Towards a Cultural Analysis of Water Use. *Current Sociology, 56*(5), 779–797.
35. Akbari, A., Nikookar, M., & Feizbahr, M., (2013). Reviewing performance of piled raft and pile group foundations under the earthquake loads. *Research in Civil and Environmental Engineering, 1*(5), 287–299.
36. Andrade, D. M., & De Freitas, R. F. B., (2022). A versatile friction model for Newtonian liquids flowing under unsteady regimes in pipes. *Meccanica, 57*, 43–72. https://doi.org/10.1007/s11012-021-01458-5.
37. Vardy, A. E., Brown, J. M. B., He, S., Ariyaratne, C., & Gorji, S., (2015). Applicability of frozen-viscosity models of unsteady wall shear stress. *J. of Hydraul. Eng., 141*, 1–13. https://doi.org/10.1061/(ASCE)HY.1943-7900.0000930.
38. Sarkardeh, H., Zarrati, A. R., Jabbari, E., & Marosi, M., (2014). Numerical simulation and analysis of flow in a reservoir in the presence of vortex. *J. Eng. Appl. Comput. Fluid Mech., 8*(4), 598–608. doi: 10.1080/19942060.2014.11083310.
39. Zhou, L., Liu, D., & Karney, B., (2013). Investigation on hydraulic transients of two entrapped air pockets in a water pipeline. *J. Hydraul. Eng.* doi: 10.1061/(ASCE)HY.1943-7900.0000750.
40. Möller, G., Detert, M., & Boes, R. M., (2015). Vortex-induced air entrainment rates at intakes. *J. Hydraul. Eng.* doi: 10.1061/(ASCE)HY.1943-7900.0001036.

41. Zecchin, A. C., Maier, H. R., Simpson, A. R., Leonard, M., & Nixon, J. B., (2005). Parametric study for an ant algorithm applied to water distribution system optimization. *IEEE Transitions on Evolutionary Computation, 9,* 175–191. doi: 10.1109/TEVC.2005.844168.
42. Ferras, D., Manso, A. P. A., Schleiss, A. J., & Covas, D. I. C., (2018). One-dimensional fluid–structure interaction models in pressurized fluid-filled pipes: A review. *Applied Science, 8*, 1844–1856. https://doi.org/10.3390/app8101844.
43. Zhao, L., Yang, Y., Wang, T., Han, W., Wu, R., Wang, P., Wang, Q., & Zhou, L., (2020). An experimental study on the water hammer with cavity collapse under multiple interruptions. *Water, 12,* 2566. https://doi.org/10.3390/w12092566.
44. Lema, M., Peña, F. L., Buchlin, J. M., Rambaud, P., & Steelant, J., (2016). Analysis of fluid hammer occurrence with phase change and column separation due to fast valve opening by means of flow visualization. *Exp. Therm. Fluid Sci., 79*, 143–153. doi: 10.1016/j.expthermflusci.2016.07.008.
45. *Minimum Design Loads for Buildings and Other Structures*, (2010). American Society of Civil Engineers.
46. Munshi, J. A., (1998). *Rectangular Concrete Tanks.* Portland Cement Association.
47. Munshi, J. A., (1993). *Circular Concrete Tanks Without Prestressing*. Portland Cement Association.
48. Munshi, J. A., (2002). *Design of Liquid Containing Concrete Structures for Earthquake Forces.* Portland Cement Association.
49. NCHRP Report 611, (2008). *Seismic Analysis and Design of Retaining Walls, Buried Structures, Slopes, and Embankments*. American Association of State Highway, National Cooperative Highway Research Program.
50. Nourbakhsh, S. A., Jaumotte, B. A., Hirsch, C., & Paris, H. B., (2008). *Turbopumps and Pumping Systems.* Springer.
51. NSF / ANSI 61, (2010). *Drinking Water System Components – Health Effects.* NSF National Standard / ANSI National Standard.
52. NZS 3106, (2009). *Design of Concrete Structures for the Storage of Liquids.* Cement and Concrete Association of New Zealand, Standards Association of New Zealand.
53. Monson, J. (2014). "Sewer Maintenance the Greeley Way– How to make the 80:20 rule work for you." *Proceedings of the Water Environment Federation, 2014*(1), 1–4.
54. Shahsavari, G., Arnaud-Fassetta, G., Bertilotti, R., Campisano, A., & Riou, F. (2015). "Bed Evolution under One-Episode Flushing in a Truck Sewer in Paris, France". World Academy of Science, Engineering and Technology, *International Journal of Civil, Environmental, Structural, Construction and Architectural Engineering, 9*(7), 814–823.
55. Stephenson, D. (1977). Effects of Air Valves and Pipe Work on Water Hammer Pressures, *J. Transp. Eng., 123*(2), 101–106.
56. TM 5–809–10, (1986). *Seismic Design Guidelines for Essential Buildings.* US Department of the ARMY, Washington, D.C.
57. USEPA, (2002). *Finished Water Storage Facilities.* U.S. Environmental Protection Agency.
58. Zecchin, A. C., Maier, H. R., Simpson, A. R., Leonard, M., & Nixon, J. B., (2005). Parametric study for an ant algorithm applied to water distribution system optimization. *IEEE Transitions on Evolutionary Computation, 9*(2), 175–191.
59. Walski, T. M., (2000). Hydraulic design of water distribution storage tanks. In: Mays, L. W., (ed.), *Water Distribution Systems Handbook.* Chapter 10. McGraw-Hill, New York.

60. Farmani, R., Walters, G., & Savic, D., (2006). Evolutionary multi-objective optimization of the design and operation of water distribution network: Total cost vs. reliability vs. water quality. *Journal of Hydroinformatics, 8*(3), 165–179.
61. Farmani, R., Savic, D. A., & Walters, G. A., (2004). *The Simultaneous Multi-Objective Optimization of Anytown Pipe Rehabilitation, Tank Sizing, Tank Siting and Pump Operation Schedules* (pp. 4663–4672). Critical transitions in water and environmental resources management, ASCE, American Society of Civil Engineers, Salt Lake City.
62. Haarhoff, J., Van, Z. J. E., Nel, D., & Van, D. W. J. J., (2000). Water supply and water quality in rural areas: Stochastic analysis of rural water supply systems. *Water Supply, 18*(1), 467–470.
63. Maier, H. R., Simpson, A. R., Zecchin, A. C., Foong, W. K., Phang, K. Y., Seah, H. Y., & Tan, C. L., (2001). Ant colony optimization for the design of water distribution systems. *Proceedings of World Water and Environmental Resources Congress*. ASCE, The Rosen Plaza Hotel, Orlando, Florida, United States.
64. Maier, H. R., Simpson, A. R., Zecchin, A. C., Foong, W. K., Phang, K. Y., Seah, H. Y., & Tan, C. L., (2003). Ant colony optimization for design of water distribution systems. *Journal of Water Resources Planning and Management, 129*(3), 200–209.
65. Walski, T. M., Chase, D. V., Savic, D. A., Grayman, W., Beckwith, S., & Koelle, E., (2003). *Advanced Water Distribution Modeling and Management*. Watertown, Conn: Haestad Press.
66. Hashemi, S. S., (2010). *Optimization of Water Networks with Minimization of Pumping Energy*. Master Thesis, University of Tehran, Iran.
67. Mau, R. E., Boulos, P. F., Clark, R. M., Grayman, W. M., Tekippe, R. J., & Trusell, R. R., (1995). Explicit mathematical models of distribution storage water quality. *ASCE Journal of Hydraulic Engineering, 121*(10), 669–709.
68. Kettler, A., & Goulter, I. (1985). "An Analysis of Pipe Breakage in Urban Water distribution Networks." *Can. J. Civ. Eng., 12*(2), 286–293.
69. Rossmann, L. A., & Grayman, W. M., (2000). Scale-model studies of mixing in drinking water storage tanks. *J. of Environmental Engineering, 125*(8), 755–761.
70. Yeung, H., (2001). Modeling of service reservoirs. *Journal of Hydroinformatics, 3*(3), 165–172.
71. Creaco, E., & Walski, T., (2017). "Economic Analysis of Pressure Control for Leakage and Pipe Burst Reduction." *Journal of Water Resources Planning and Management, 143*(12).
72. Mahmood, F., Pimblett, J. G., Grace, N. O., & Grayman, W. M., (2005). Evaluation of water mixing characteristics in distribution system storage tanks. *J. AWWA, 97*(3), 74–88.
73. Marek, M., Stoesser, T., Roberts, P. J. W., Weitbrecht, V., & Jirka, G. H., (2007). *CFD Modeling of Turbulent Jet Mixing in a Water Storage Tank*. XXXII IAHR Congress, Venice, Italy.
74. Van, Z. J. E., & Haarhoff, J., (2007). Reliability analysis of municipal storage reservoirs using stochastic analysis. *Journal of the South African Institution of Civil Engineering, 49*(3), 27–32.
75. Gondo, R., & Kolawole, O. D. (2019). Sustainable Water Resources Management: Issues and Principles of Water Governance in the Okavango Delta, Botswana. *International Journal of Rural Management, 15*(2), 198–217.
76. Heino, O., & Takala, A. (2020). Transformation of Urban Water Service Provision: Potential of Hybrid Systems. *Public Works Management & Policy, 25*(2), 151–166.

CHAPTER 2

Biopolymer Materials: Sensors, Energy Devices, and Storage Applications

NARENDER SINGH, SONIA KHANNA, and ANJIKA BOORA

Advanced Polymeric Materials Research Laboratory, Department of Chemistry and Biochemistry, Sharda University, Greater Noida, Uttar Pradesh, India

ABSTRACT

Biopolymers having high biocompatibility, biodegradable factors, and several active sites are commonly used in biomedical applications. Currently, the unique electronic abilities of biopolymers are being used to create wearable, implantable, and biodegradable electronics. This chapter provides a broad overview of advanced biopolymers related to energy, sensors, displays, and bionic devices in the context of molecular engineering, including molecular design and assembly, structural modification, and highlights the important roles of structural modification in tuning device performance. Furthermore, typical printing processes are explored to emphasize their advantages in the low-cost, eco-friendly, and scalable fabrication of biopolymer electronics. Additionally, the existing problems and future potential for electronic biopolymers for the next generation of flexible, high-strength, and wearable bioelectronics are anticipated.

2.1 INTRODUCTION

2.1.1 WHAT ARE BIOPOLYMERS?

Polymeric biomolecules with a long chain of monomeric units that are covalently joined together are known as biopolymers [1]. Biopolymers, unlike

Sustainable Environmental Engineering: Water Security, Energy Conservation, and Green Process.
Sonia Khanna, Ann Rose Abraham, Kian Hariri Asli, & A. K. Haghi (Eds.)

synthetic polymers, have a complex molecular arrangement that results in a well-defined three-dimensional structure. The term "bio" refers to the polymer's biological ability to degrade naturally. They have the ability to mimic the characteristics of synthetic polymeric materials [2]. Biopolymers are a very interesting alternative for extremely sensitive and selective gas and vapor phase sensors because of their widespread use, biocompatibility, and unique properties. Biopolymer-derived materials are renewable and sustainable, with excellent performance and a low carbon footprint [3]. Biopolymers' exceptional properties make them an ideal option for industrial wastewater treatment. Biopolymers and their derivatives have the potential to be a long-term replacement for petroleum-derived wastewater materials. Biopolymers are a diverse and adaptable type of material with potential applications in almost every sector of the economy. They may be utilized as adhesives, absorbents, lubricants, soil conditioners, cosmetics [4], drug utility and delivery vehicles, textiles materials, high-strength structural materials, and even devices related to computational analysis. Many biopolymers are still in the early stages of development, but significant applications in packaging [5], food production [6], and medicine are beginning to emerge. Some biopolymers can directly replace synthetically obtained materials in traditional applications as well as routes, while others have unique properties that could lead to a plethora of new commercial opportunities.

2.1.2 CLASSIFICATION OF BIOPOLYMERS

Various scales can be used to categorize biopolymers [7]. These divisions are made in accordance with their origin, the quantity of monomeric units, degradability, heat response, etc., as shown in Figure 2.1. Some of them are explained below with examples and their applications.

2.1.2.1 CELLULOSE

It is the first thermoplastic polymer found in plants having the chemical formula $(C_6H_{10}O_5)_n$ where n is the number of repeating units present [8]. The IUPAC name for cellulose is (6,5)-2-(hydroxymethyl)-6-(3S-4,5,6-trihydroxy-2-(hydroxymethyl)oxan-3-yl)oxyoxan-3,4,5-triol. It comes under the polysaccharide with a linear structure of 100–1,000 repeating units of (1-4) β-linked glucose units. D-Glucose is the glucose unit that has (1-5) OH and CH_2OH groups in the same plane as shown in Figure 2.2. Cellulose is a homopolymer with no toxicity and decomposable behavior. The tensile and compressive strength of cellulose [9] is outstanding as they consist of cross-linked structure.

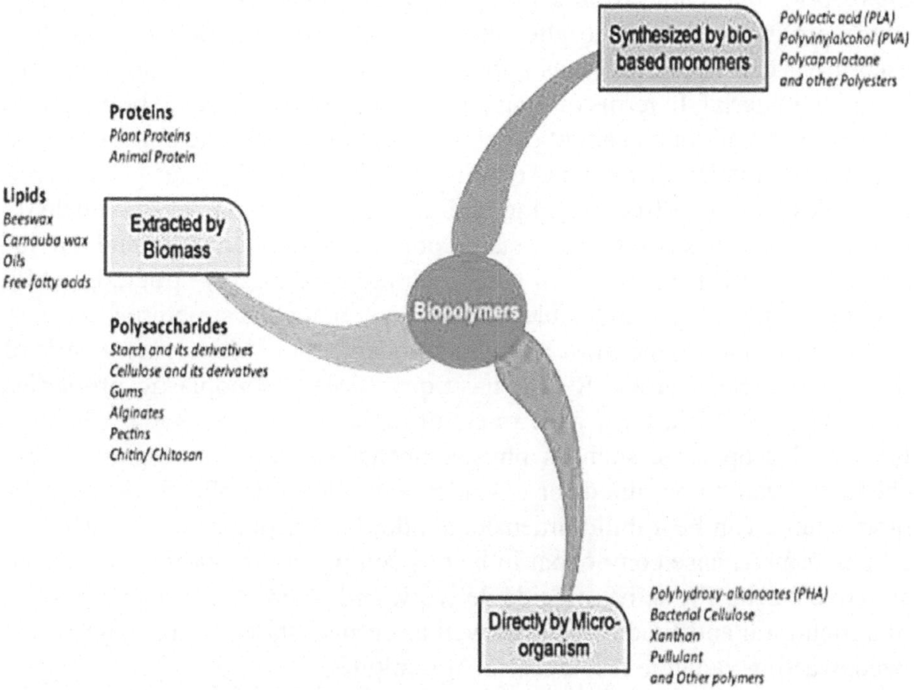

FIGURE 2.1 Classification of biopolymers.

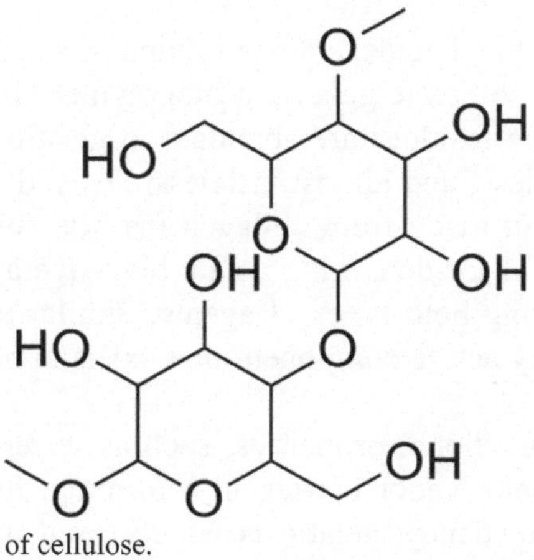

FIGURE 2.2 Structure of cellulose.

2.1.2.2 STARCH

The main source of energy is starch. It makes up 60 to 75% of the weight of grain products. For consumption, it requires between 70% and 80% of the calories that come from a living thing or human body [10]. Typically, starch products are used to alter the physical characteristics of food. It is used in various applications as a thickening agent, adhesive, and moisture-retaining material. In terms of structure, starch and cellulose are both natural biopolymers with only a slight difference in internal bonding between them. Starch contains D-glucopyranose units that are linked together [11]. Starch molecules have one free reducing end due to the presence of the aldehyde group. Glucose has two types of starch polymerization: amylose and amylopectin. Amylose is an essential carbohydrate with a linear structure, while amylopectin is a larger molecule with a highly branched structure [12].

Starch is among the most abundant biodegradable biopolymers derived from renewable sources. It also has convertible thermoplastic properties that make it suitable for a wide range of agricultural applications. Starch's functional properties, such as physicomechanical, boundary, and surface chemistry, can be modified for extended agricultural applications. Furthermore, starch can be a multidimensional additive for plasticulture, acting as a filler, a metaphase component in blends/composites, a plasticizer [13], an effective carrier for active biocide delivery, and so on. A significant portion of agricultural and food wastes, as well as surplus starch sources, go unutilized, wasting valuable resources for agriculture.

2.1.2.3 GELATIN

In the presence of diluted acids, collagen found in animal skin and bones is thermally denatured to create gelatin, a biopolymer [14]. Glycine, proline, and 4-hydroxyproline residues are abundant in gelatin. It is a powder that is translucent, colorless, and almost tasteless. Depending on the extraction technique, the gelatin made from collagen may be acidic or basic. While basic gelatin is useful for delivering acidic bioactive agents, acidic gelatin is useful for delivering both types of agents. Similar to any other protein, gelatin is an optically active component that exists in both random coil and helical states [15].

Because of its functional properties, such as water-binding ability, gel formation ability, water vapor barrier, film forming, foam forming ability, and emulsification tendency, gelatin is widely used in the food, pharmaceutical [16], photography, and cosmetics industries. Gelatin has good gas barrier properties and swelling behavior in water, but it has poor mechanical resistance and permeability to water vapor molecules.

2.1.2.4 CHITOSAN

Cellulose, Dextrin, Agar, and Starch, etc., are acidic natural biopolymers, but chitosan is the basic biopolymer [17]. Solubility, viscosity, and polyelectrolyte behavior are among its properties. It is capable of forming films and producing polyoxy salt. Chitosan can be prepared by using shrimp shells as shown in Figure 2.3.

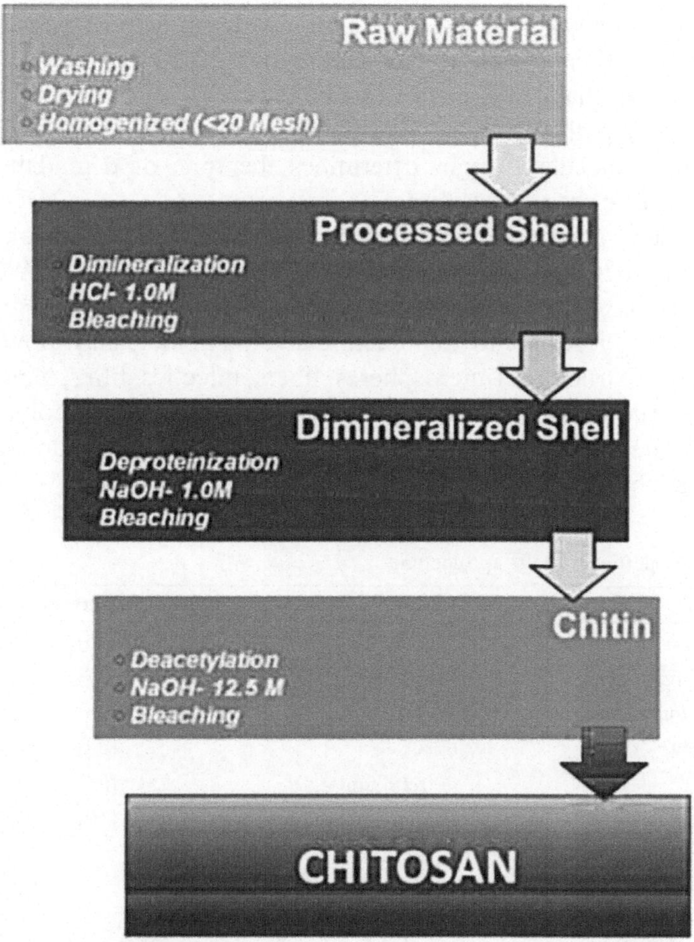

FIGURE 2.3 Chitosan obtained from shrimp shell.

Just like cellulose, it is insoluble in water and some organic solvents [18]. It is soluble in mineral acid and dimethylacetamide aqueous solutions [19]. Chitosan is hydrolyzed with concentrated mineral acids under harsh conditions to produce pure sugar and D-glucosamine [20]. It generates deacylated chitin and cationic polysaccharide derivatives. When it reacts with polysaccharides, it tends to form a polyelectrolyte complex [21].

2.1.2.5 POLYCAPROLACTONE

Polycaprolactone is easily degradable with a biocompatibility factor and is a bioresorbable polymer that is an aliphatic polyester belonging to the group of polyhydroxy acids, which includes polylactic and polyglycolic acids. Firstly, it was synthesized in the early 1930s with the help of ring-opening polymerization of Ɛ-caprolactone cyclic monomer [22, 23]. Polycaprolactone is composed of a chain of Ɛ-caprolactone $(C_6H_{10}O_2)_n$ repeating single-unit sequences. The length (n) of the Polycaprolactone chain or the polymer's corresponding molecular weight determines the time of degradation and persistence via ester-bond hydrolysis.

PCL is a semi-crystalline, hydrophobic polymer with a glass transition temperature of 60?C and a low melting point varying from 59 to 64?C [24]. It additionally has better viscoelastic properties than other biodegradable polymers, making it simple to manufacture and manipulate and allowing for a wide variety of structures (microspheres, fibers, micelles, films, nanofibers, foams, etc.). Additionally, it is simple to combine with other polymers to create copolymers, which have a variety of physicochemical characteristics and biodegradability [25] (Table 2.1).

TABLE 2.1 Properties of Polycaprolactone

Properties	Values	References
Density	1.146g/mL at25°C	[26]
Molecular Weight	Mn~10000 by GPC	[27]
Glass Transition Temperature	Tg-60°C	[28]
Melting Point	60°C	[28]
Viscosity	400-1000mPa.s	[29]
From	Flakes or Chunks	[30]

2.1.2.6 POLY(VINYL ALCOHOL)

The hydrolysis of its precursor polyvinyl acetate produces PVA, a thermoplastic biopolymer as shown in Figure 2.4 [31]. The degradation process is carried out by biological microorganisms [32, 33]. Due to its higher crystallinity, it is highly soluble in water. Many polymer end products, including alcoholic beverages, surgical threads, food packaging, etc., are made using PVA [34, 35]. It is a polymer that is robust, ductile, and remarkably flexible [36].

Table 2.2 shows the properties of poly(vinyl alcohol) which vary depending on the grade, molecular weight, and degree of hydrolysis [37]. It possesses exceptional flexibility, tensile strength, and gas barrier qualities [38]. It is a very popular biopolymer that is easily kept on the surface of water and can chemically bond to the surface of materials. Poly(vinyl alcohol) is a toxic-free and water-soluble biopolymer and it has many industrial uses as pointed out below:

- Packaging in food industries.
- In paper making industries, it improves the durability of the paper and enhances the quality of the paper [39].
- Used in the textile industry, providing a lot of products with different variations and patterns.
- In the building and construction sector, its binding capability improves the construction material [40].
- It is used in the manufacturing of various personal and household products, making them safe and hygienic.
- In biomedical industries, it is used in the production of biomedical products such as contact lenses, surgical threads, and many other tools used in the biomedical field.

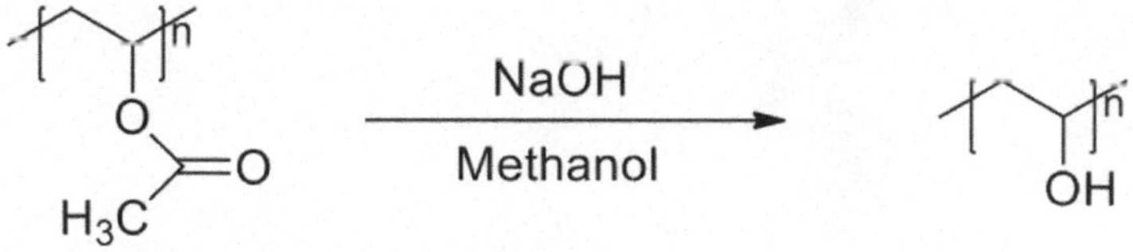

FIGURE 2.4 Synthesis of poly(vinyl alcohol).

TABLE 2.2 Properties of Poly(Vinyl Alcohol)

Properties	Values	References
Density	1.20g/cm^3	[41]
Flash point	352.59K	[41]
Glass Transition Temperature	400	[42]
Melting Point	473K	[43]
pH	5.0-7.0	[44]
Refractive Index	1.46 at 631nm	[44]

2.1.2.7 PROTEINS

Proteins are vital components of daily life for human beings and other living beings on this planet. They are one of the essential components that make up body tissue. Peptide bonds are used to link the amino acid components which make up the protein [45]. Peptide linkages are those linkages which are created by amino acid and carboxylic acid molecules [46]. A polypeptide bond holds the lengthy macromolecular chain of proteins together [47]. All biological processes rely heavily on proteins. They serve as catalysts, transport other molecules from one place to another, or store them (oxygen). Moreover, they support the immune system mechanically. Proteins are present throughout your body, i.e., in muscles, skin, hair, and almost every body part and tissue.

1. **Keratin:** Protein polymer known as keratin is obtained from molecules modified in the skin of animals such as horns, claws, and hooves [48]. It consists of a variety of floating proteins, enzymes, and keratinized filaments. It is described as a specific filament-floating protein with outstanding physiochemical characteristics. Layers of cellular epidermis are used to extract it [49]. Figure 2.5 shows the structure of the protein.

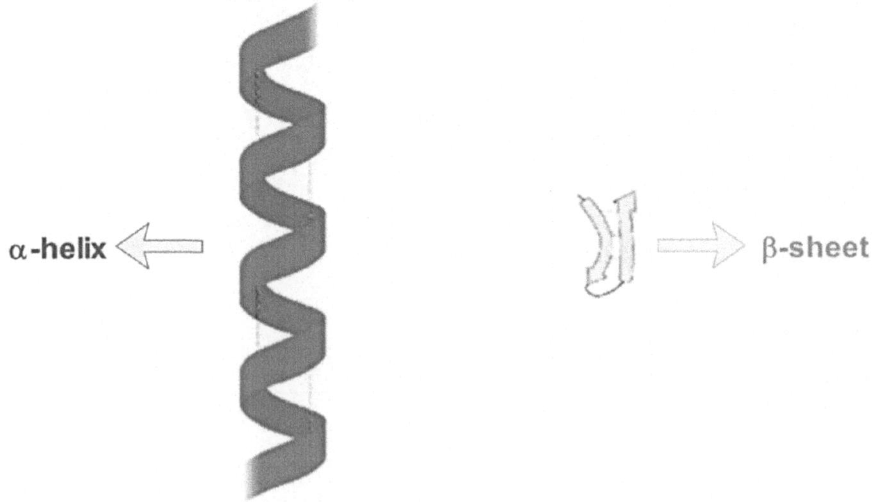

FIGURE 2.5 The structure for protein.

Silk is one of the protein fibers which is produced by silkworms, spiders, mites, fleas, and scorpions. For the creation of biomaterials, proteins taken from spider silk, Bombyx mori or mulberry silk, as well as non-mulberry silks, have been investigated. For the creation of a spider web, spider silk from the major ampullate glands (MA silk), draglines, and flabelliform is essential. Depending on the type of silk, oligopeptides are arranged structurally.

2. **Fibrin:** The main element of the coagulation cascade is fibrin, one of the fibrous proteins. After tissue injury, fibrin is an insoluble protein created from soluble fibrinogen to stop the bleeding. The creation of insoluble fibrin matrix from its soluble precursor fibrinogen by the action of one serine protease: the thrombin, at the site of injury, starts the coagulation cascade [50]. A pair of A, B, and C chains connected by a disulfide bond make up fibrinogen [51]. Moreover, it has a core globular region (E domain), two distal globular regions (D domain), and a helical coiled coil connecting them. The nodules that make up the E domain are joined by non-covalent contact.

   Chrobak et al. layered the fibrin micro-threads into the fibrin hydro gel to create a mechanically stable biomaterial for cardiac tissue healing. The made-up fibrin hydro gel with integrated micro-threads of fibrin demonstrated mechanical qualities resembling those of cardiac tissue. Moreover, the created biomaterial aids in cell contraction and orientation [52].

2.1.2.8 LIPIDS

Lipids are materials obtained from plants and animals by dissolving them in either chloroform or methanol [53]. When the parts of plants and animals are pulverized and dissolved in non-polar solvents, lipids are separated. Lipids are the only non-polar compounds produced by this process. They are not only identified by a single structure but also by the method used to isolate them. In nonpolar liquids, proteins, and hydrocarbons do not dissolve [54]. Lipids, one of the three major types of food ingredients, are a source of energy. Because they are concentrated in a tiny space and have substantially reduced carbon atoms on their acyl chains, lipids store energy even more effectively than carbohydrates. Lipids can be classified into three categories: simple, compound, and derived lipids (Table 2.3).

TABLE 2.3 Classification of Lipids with Sources

Types of Lipids	Source	Examples
Simple Lipids	Vegetable Oil	Alomnd, Apricot,Walnut
	Animal Oil	Black Sea dogfish, Emu, Sardine
	Saturated fats	Caprylic, Lauric
	Unsaturated fats	Oleic, Linolenic
Compound Lipids	Phosphatidyl glycerol	1,2-Dimyristoyl-sn-glycero-3-Phosphoglycerol (DMPD)
	Diphosphatidyl glycerol	
	Phosphatidyl cholines	1,2-Dimyristoyl-sn-glycero-3-Phosphocholines (DMPD)
	Sphingophospholipids	Sphingomielins
	Sphingoglycolipids	Cerebrosides
Derived Lipids	Sterols and sterol esters	Cholestrols
	Vitamin A, D, E and K	Carotenoids,Retinol, Cholecalciferol, Tocopherols,Warfarin

2.2 RECENT ADVANCEMENTS IN BIOPOLYMER BASED-MATERIALS

The demand for energy is growing day by day, while the supply of fossil fuels, which currently meet nearly 80% of the world's energy demands, is depleting. The most popular non-renewable energy sources are expected to be depleted by 2200 (oil-based energy will last for 40 years, natural gas for 60 years, and coal for 200 years) [55]. Therefore, considering different sustainable energy sources should be among our top priorities. In the base-case forecast, the total global demand for biofuels is expected to increase by 20% between 2022 and 2027. Almost all of the growth in the consumption of renewable diesel and biojet fuel is expected to occur in developed economies. Policies intended to reduce GHG emissions are driving demand for these fuels, as they can be produced with low GHG emissions [56], blended at high levels, and made from wastes and residues. In fact, in 2021, wastes and residues provided nearly 70% of the renewable diesel and biojet fuel. Around 80% of the world's growth in the use of biofuels comes from the United States, Canada, Brazil, Indonesia, and India, all of which have comprehensive growth-supporting policy frameworks [57]. In Europe, even though state-level regulations are becoming more difficult, declining transportation fuel demand almost completely halts volume growth. From 2022 to 2027, the percentage of biofuels in the world's transportation fuel consumption is expected to increase from 4.3% to 5.4%. To overcome all energy-related issues, researchers have already started working on the source of energy from biopolymer-based materials, as they are new, advanced, and cost-effective alternatives to other materials in terms of cost [58], productivity, concerned modifications, strength factor, durability factor, effect on the environment, and many more [59].

The growth of electrochemical storage [60] mechanisms for energy from copious, renewable, eco-friendly, and cost-effective materials has been a point of focus and driving force in the advancement of electronic devices. This task and demand can be met with an abundant resource that has the potential to be developed into effective electrochemical energy storage systems. Lignin and cellulose, also known as lignocellulose, are two of the most prevalent biopolymers, and they have the attributes and qualities to meet this demand [61]. Because of their abundance and diversity of functional groups, lignocellulose biopolymers have unique properties such as mechanical flexibility, porosity, and convertibility. These characteristics make it possible to combine them effectively with electronically conducting polymers as improved materials for the creation of long-term energy storage technologies [62]. The difficulties lie in the use of lignocellulose with electronically conducting polymers in various capacities as electrodes, separators, binders, and electrolytes in supercapacitor applications and battery applications [63] (Table 2.4).

TABLE 2.4 Types of Biopolymers Used in Global Commercial Market

Biopolymers	Major Producers	Application	Biodegradable
Collagen and gelatin	Integra LifeSciences,Royal DSM,Collplant Biotechnologies	Orthopedic,Wound care Dental	Yes
Polylactic acid (PLA)	Nature Works,Evonik,Total Cordion PLA	Flexible Packaging, Rigid packaging, Consumer good	Yes
Polyhydroxyalkanoates (PHA)	Yield 10 Bioscience,Tianjin GreenBio Material,Bio-on	Flexible Packaging, Rigid packaging	Yes
Polyethylene (PE)	Neste,LyondellBasell	Flexible Packaging,	NO
Polyethylene terephthalate(PET)	Toray Industries,The Coca-Cola Company,M&G Chemicals	Rigid packaging	NO
Polybutylene adipate terephthalate (PBAT)	Algix,BASF	Flexible Packaging, Rigid packaging agriculture	Yes
Polybutylene succinate (PBS)	Roqutte,Mitsubishi Chem,Succinity	Flexible Packaging, agriculture	Yes

TABLE 2.4 *(Continued)*

Biopolymers	Major Producers	Application	Biodegradable
Starch and blends	Futerro,Novamont, Biome	Flexible Packaging, Consumer good, agriculture	Yes
Corn and Cotton	Corn Ware,Natureworks LLC,Klockner Pentaplast	Packaging Material,Food Bags	Yes

Triboelectric nanogenerators (TENG) have recently made strides in improving power generation and conversion efficiency [64]. However, barriers to the economical and biocompatible use of TENGs still predominate. Chitosan, an abundant natural biopolymer [65] derived from the shells of marine crustaceans, opens up promising possibilities for inexpensive, biodegradable TENG applications in related fields. Here, for the first time, is the development of flexible, biodegradable TENGs based on chitosan. In order to optimize triboelectric power generation, the physical and chemical properties of the chitosan nanocomposites [66] are systematically studied and engineered, turning otherwise useless natural materials into useful energy devices as shown in Figure 2.6 [64–67].

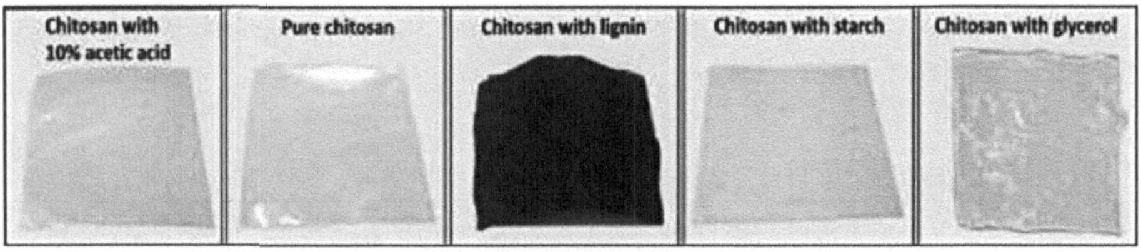

FIGURE 2.6 Nanocomposites made from chitosan for TENG.

Studies on polymer electrolytes now include biopolymers as a key component. A membrane made of organic or inorganic salts that have been dissolved in a polymer is known as a polymer electrolyte [68]. Electron donor atoms like N, O, or S atoms must be present in the polymer that serves as the host for ionic conduction. These atoms barely interact with the dissolved salts' cations. Both cations and anions contribute to conductivity in polymer electrolytes [69]. Comparing polymer electrolytes to their liquid counterparts reveals several advantages. Polymer electrolytes can solve the leakage and corrosion issues that are related to conventional liquid electrolytes. Polymer electrolytes also benefit from lighter weight devices, flexibility in design, and the ability to act as separators between electrodes [70–72]. Since they are plentiful, affordable, and environmentally friendly, polysaccharides among the biopolymers have a good chance of succeeding as hosts for ionic conduction and could eventually replace synthetic polymers in energy production and storage devices. One such substance is chitosan. Chitin can be made into it by deacetylating it until it can dissolve in weak acids. Another illustration is methyl cellulose [73]. It can be made by methylating the alkaline cellulose after treating it with caustic soda. The chitosan and cellulose-based materials can be used in energy generation and devices.

2.2.1 DYE-SENSITIZED SOLAR CELLS

Since its first construction by Brian and Grätzel in 1991 [74], the dye-sensitized solar cell (DSSC) has garnered considerable attention as one of the most promising methods for converting solar energy [75]. The operating system of this third-generation solar cell shares many similarities with how plants synthesize photosynthetically in nature [76]. This technology is very appealing due to its ease of fabrication, low cost of raw materials, processing techniques using basic, reasonably priced equipment, use of environmentally friendly components, and ability to create flexible solar cells [77]. One of the essential parts of dye-sensitized solar cells is the electrolyte material, which transports inner charge carriers between the photo-anode and the counter electrode and constantly regenerates the dye while in operation. They have a significant impact on the long-term stability and light-to-electric conversion efficiency of dye-sensitized solar cells [78]. A variety of ionic liquids, liquid electrolytes based on metal salts, and redox couples have been used in dye-sensitized solar cells since the beginning.

In the previous two decades, there has been a lot of study into the development of biopolymer electrolytes for dye-sensitized solar cells. However, this area lacks comprehensive information. In a recent study, Singh and co-workers highlighted the current findings from laboratory investigations and discussed the potential of solid-state biopolymer electrolytes in dye-sensitized solar cells and battery applications [79]. Thus, there is a great need for a thorough critical assessment of the biopolymer electrolytes in the context of dye-sensitized solar cell applications.

The dye-sensitized solar cell has a number of benefits over other solar cells, including lower manufacturing and maintenance costs, a smaller weight, and several customizable processing methods. Furthermore, unlike first- and second-generation solar cells [80, 81], some dye-sensitized solar cells can perform better even in dim lighting conditions (such as cloudy, rainy, or dusk). The creation of effective and commercially viable dye-sensitized solar cells, such as lightweight, roll-to-roll or sheet-like structured dye-sensitized solar cells, has recently been the subject of research. However, the majority of these dye-sensitized solar cells generally have five parts: transparent conductive oxide (TCO), semiconductor metal oxide [82] or photoanode, dye sensitizer, electrolyte, and counter electrode. The semiconductor metal oxide (photoanode) of DSSCs is covalently attached to a monolayer of dye molecules, also known as a dye sensitizer. The dye sensitizer interacts physically and electrically with electrolytes and semiconductor materials. At the dye/SMO contact, it releases photoexcited electrons that are then sent to the TCO after it absorbs light via solar radiation [83, 84], as shown in Figure 2.7.

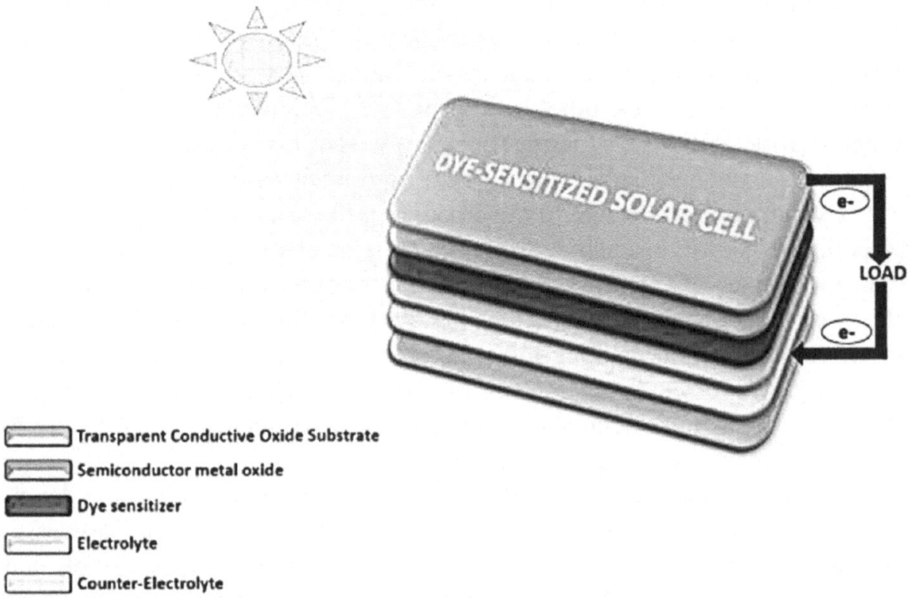

FIGURE 2.7 Schematic representation of dye sensitized solar cell.

2.2.1.1 BIOPOLYMER-BASED ELECTROLYTES BY CHITOSAN AND ITS DERIVATIVES

Chitosan is a linear polysaccharide made of D-glucosamine that is -(1-4) linked. The acetylated D-glucosamine remains randomly distributed among the deacetylated units because deacetylation is never complete [85]. Chitosan is created by soaking the shells of prawns and other crustaceans in a sodium hydroxide solution. Figure 2.8 represents the structure of chitosan.

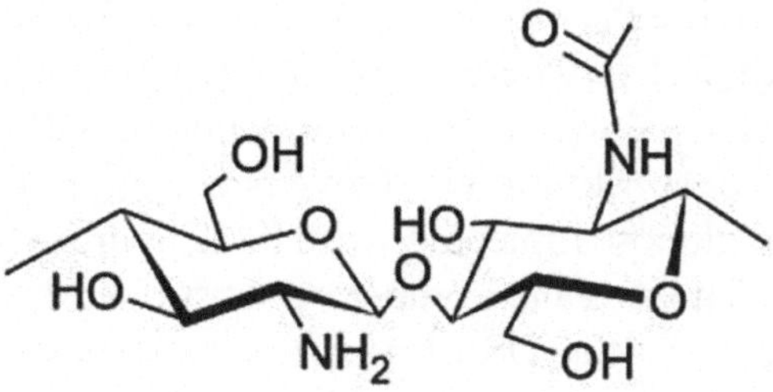

FIGURE 2.8 Structure of chitosan.

The conductivity of a chitosan film created by combining chitosan powder with 1% acetic acid has low conductivity at room temperature (between 10^{-10} and 10^{-9} S cm^{-1}) [86]. However, it can solvate many inorganic salts and form good films, acting as a matrix for ionic conduction. When chitosan is solvated with lithium salt or ammonium salt for proton conduction, the conductivity at room temperature improves. Some chitosan-based lithium ion and proton conductors are listed in Table 2.5. The plasticized electrolytes exhibit high room temperature conductivity, as shown in Table 2.5. This is because ionic conduction in polymer electrolytes occurs in the amorphous region, and plasticization can increase the electrolyte's amorphous content [87].

TABLE 2.5 Chitosan and Its Derivative Ionic Conductivity at Room Temperature

Material with Composition	Conductivity at Room Temperature	References
Chitosan – 40 wt.% oxalic acid	4.95×10^{-7}	[88]
70 wt.% phthaloyl chitosan – 30 wt.% NH_4SCN	$(2.42 \pm 0.01) \times 10^{-5}$	[89]
37.50 wt.% chitosan – 37.50 wt.% iotacarrageenan – 18.75 wt.% H_3PO_4 – 6.25 wt.% PEG.	6.29×10^{-4}	[90]
60 wt.% (Chitosan: PEO wt.% ratio 3:2) – 40 wt.% NH_4NO_3	1.02×10^{-1}	[91]
30 wt.% (36 wt.% PVA – 24 wt.% chitosan – 40 wt.% NH_4NO_3) – 70 wt.% EC	1.60×10^{-3}	[92]

2.2.1.2 BIOPOLYMER-BASED ELECTROLYTE BY CELLULOSE AND ITS DERIVATIVES

By heating cellulose in an alkaline solution (such as one containing sodium hydroxide) and treating it with methyl chloride, methyl cellulose (MC) can be derived synthetically [93]. Methoxide replaces the hydroxyl (-OH)

functional groups during the reaction (–OCH_3 groups). The chemical formula is $[C_6H_7O_2(OH)_x(OCH_3)_y]_n$, where 'x' can range between 1.00 and 1.55 and y between 2.00 and 1.45, with a maximum value of 3.00 for the sum of 'x' and 'y.' The degree of substitution is shown here (DS) [94]. A polymer made from several linked glucose molecules with three hydroxyl functional groups is called cellulose. The average number of substituted hydroxyl groups per glucose is referred to as the DS of a particular form of methyl cellulose. Depending on how it was made, the MC backbone's average length can vary. In the structure of methyl cellulose, R can be either an H group or a CH_3 group as shown in Figure 2.9, and Table 2.6 provides some of the salts dissolved in MC as well as their conductivity values.

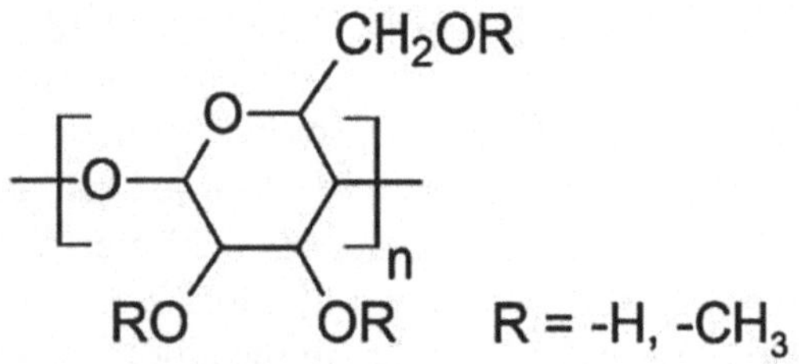

FIGURE 2.9 Structure of methyl cellulose.

TABLE 2.6 Ionic Conductivities of Methyl-Cellulose-based Electrolytes at Room Temperature

Material with Composition	Conductivity at Room Temperature	References
85 wt.% (methyl cellulose – 30 wt.% KOH)–15 wt.% dimethyl carbonate	~10^{-5}	[95]
75 wt.% methyl cellulose – 25 wt.% NH_4NO_3	2.10×10^{-5}	[96]
91 wt.% methyl cellulose – 9 wt.% LiBOB	~10^{-7}	[97]
70 wt.% methyl cellulose – 30 wt.% $LiCF_3SO_3$	2.13×10^{-5}	[98]
Hydroxypropyl methyl cellulose: $CdCl_2$ in ratio (5:4)	1.45×10^{-6}	[99]

2.2.1.3 BIOPOLYMER-BASED ELECTROLYTE BY AGAROSE

Agar is a seaweed-derived linear polysaccharide made up of alternating -D-galactose and -1/4-linked 3,6-anhydro-L-galactose units [100]. It is commonly employed as an aqueous electrolyte gelator. Gelation temperature can be controlled by using methods like methylation, alkylation, and hydroxyalkylation of the polymer. In warm water and other organic solvents like DMF, DMSO, DMA, NMP, etc., it is readily soluble; however, it has very poor solubility in cold water [101]. It can create a robust, cross-linked matrix that supports electrolytes and facilitates the transfer of mobile ions without impairing the liquid's conductivity. Its conductivity ranges from (10^{-2} to 10^{-3} S/cm), making agarose-based electrolyte a promising polymer matrix for dye-sensitized solar cell [102] construction. Several agarose-based electrolyte DSSCs' photovoltaic capabilities are tabulated in Table 2.7. In general, it is clear that agarose-based electrolytes performed better in terms of conductivity than cellulose-based electrolytes. The highest efficiency, meanwhile, was just about 6%, which is less effective than electrolytes based on cellulose. That might be explained by the fact that efficacy is not solely dependent on conductivity. Moreover, pure agarose-based electrolytes have low stability, which is possibly due to agarose's extremely poor heat insulating ability. Although some of this problem has been solved by the incorporation of nanoparticles, the effectiveness of these NP-incorporated electrolytes is poor.

TABLE 2.7 Photovoltaic Parameters for Agarose-based Polymer Electrolyte

Electrolyte	Solvent	σ (mS/cm)	Dye	D (μm)	JSC (mA/cm²)	VOC (V)	FF	References
agar/MPII/LiI/I_2	MPN	–	N719	30	6.77	0.650	0.67	[103]
agar/MPII/TiO_2/I_2	NMP	–	–	–	10.96	0.545	0.570	[104]
agar/KI/I_2	Water	9.02	–	–	3.27	0.67	0.24	[105]
agar/KI/I_2	DMSO	–	N3	–	0.001	0.57	0.56	[106]
agar/KI/I_2	DMF	~10	N3	–	0.001	0.75	0.77	[106]
agar/EMDCA/KI/I_2	DMSO	7.41	N3	–	0.0019	0.63	0.60	[106]
agar/EMDCA/KI/I_2	DMF	42.3	N3	–	0.0033	0.62	0.66	[106]
agar/MPII/I_2	PC+DMSO	–	N719	60	11.73	0.70	0.64	[107]
agar/AEII/I_2	PC+DMSO	–	N719	60	11.71	0.62	0.65	[107]
agar/APII/I_2	PC+DMSO	–	N719	60	11.53	0.70	0.62	[107]
agar/DAII/I_2	PC+DMSO	–	N719	60	11.84	0.70	0.60	[107]
agar/TiO_2/LiI/I_2	NMP	2.66	N719	–	5.28	0.605	0.55	[108]

2.2.1.4 BIOPOLYMER-BASED ELECTROLYTE BY STARCH AND ITS DERIVATIVES

Since starch-based electrolytes displayed mechanical stability and offered minimal steric hindrance for the mobile ions, starch, a carbohydrate biopolymer, has attracted considerable attention for use as an electrolyte for dye-sensitized solar cells [109]. The amylose content affects both the stability and the ease of ion transport. As a result, DSSC electrolyte manufacture frequently uses starches with a high amylose concentration, such as rice starch (RS), potato starch (PS), arrowroot starch, and sago starch [110]. Khanmirzaei and his team have researched the potential use of RS with a high amylose content and low viscosity as an electrolyte for DSSC production in great detail [111, 112]. They created a polymer electrolyte for DSSC application using biodegradable rice starch and LiI solution in a study. The results of this investigation demonstrated that adding LiI salt to the composite made it more amorphous, which boosted conductivity. By adding 35 wt.% of LiI to the mixture, the conductivity of the rice starch-based electrolyte was increased from 6.87×10^{-10} to 4.87×10^{-5} S/cm [113]. Eventually, they created a nanocomposite using rice starch, LiI, ionic liquid 1-methyl-3-propylimidazolium iodide (MPII), and TiO_2 nanopowder [114]. The conductivity of the aforementioned composite was improved by about eight times with the addition of nano-TiO_2 and MPII. Under one sun irradiation, this nanocomposite-based DSSC had an efficiency of 0.17%.

Several chemical modification methods, such as grafting, esterification, or oxidation, can be employed to improve thermoplastic and thermal stability while decreasing viscosity and crystalline properties. Lobregas and colleagues produced a higher amorphous modified cation potato starch by grafting 1-glycidyl-3-methylimidazolium chloride (GMIC) on the backbone of raw potato starch (CPS). In the presence of a KI/I_2 redox couple and a GMIC ionic liquid plasticizer, this CPS was water-soluble and formed a gel-like electrolyte in DMSO. Compared to its liquid electrolyte DSSC, the electrolyte demonstrated the best conductivity of 0.318 mS/cm and the maximum efficiency of 0.63% with a higher fill factor. Despite its low performance, the DSSC of the CPS electrolyte retained a stability of 42% after 20 days, which was greater than the GMIC liquid electrolyte due to superior electrode filling [115].

To avoid the need for water as a solvent in starch-based electrolytes, phthaloyl rice starch (PhRS), an organo-soluble esterified starch derivative polymer, has recently been employed in the manufacture of DSSC electrolytes. PhRS was combined with hydroxyethyl cellulose (HEC) to form a gel that contained TPAI/I_2 as a redox mediator and DMF as a solvent. The produced gel demonstrated improved mechanical strength and conductivity when compared to pristine starch-based electrolytes. Under 1 sun irradiation, a combination of 70% PhRS and 30% HEC gel electrolyte demonstrated 3.02% efficiency and 9.02 mA/cm^2 current density. Its efficiency was raised to 3.94% by increasing the TPAI concentration from 5% to 12.5%, with a maximum efficiency of 3.67%. The LiI/I_2 redox mediator could be synthesized using DSSC using a similar polymer blend gel [116, 117].

2.2.1.5 BIOPOLYMER-BASED ELECTROLYTE BY CARRAGEENAN AND ITS DERIVATIVES

Carrageenan is a class of linear sulfated polysaccharides derived from red edible seaweeds [118] that include D-galactose and 3,6-anhydro-D-galactose. Carrageenans are classified into three varieties based on the number and location of the ester sulfate groups: iota (iota) carrageenan ($C_{24}H_{34}O_{31}S_{44}$), kappa (kappa) carrageenan ($C_{24}H_{36}O_{25}S_{22}$), and lambda (lambda) carrageenan ($C_{12}H_{19}O_{20}S_{33}$) [119, 120]. Its water solubility and gelation qualities make it suitable for use in the food and pharmaceutical industries [121, 122]. This polymer works as an anionic polyelectrolyte due to the presence of an SO_3^- ion group in its structure, and from the standpoint of solid-state electrochemistry, this polymer can operate as a great polymer host by offering many coordination sites for metal cations for transport. The oxygen atoms in the polymer operate as a transporter for the charged salt species and can establish covalent connections with the cations of the redox mediators utilized in the cells [123]. Furthermore, the presence of a significant number of hydroxyl groups aids carrageenan in forming crosslinks with other electrolyte components [124, 125].

Bella and her co-workers also found that carboxymethyl-carrageenan (CMKC) has two orders of magnitude higher conductivity than pure carrageenan. They discovered that the electrolyte with 40% NaI and 30% EC had the maximum conductivity of 3.25 mS/cm. The efficiency of DSSCs made with this improved electrolyte was 2.08%, and it was proven to be stable in aging tests conducted during thermal stress [126].

Shamsudin and his team improved the conductivity of the electrolyte system by replacing NaI with ionic liquid in CMKC. The gel electrolyte included 30 wt.% 1-butyl-3-methylimidazolium chloride (BMIC) ionic liquid, which resulted in a conductivity of 5.76 mS/cm. By carboxymethylation, the conductivity and efficiency of hybrid/-carrageenan and -carrageenan based electrolytes can also be improved. Apart from increasing the amount of oxygen available for ion transport, carboxymethylation lowers the crystalline structure of the gel, which improves polymer segmental mobility [127, 128].

2.2.2 BATTERIES

2.2.2.1 LITHIUM-SULFUR (LI-S) BATTERIES

Lithium-sulfur batteries (Li-S) are good candidates for large-scale energy storage applications due to their high theoretical specific capacity (1,672 mAh g^{-1}), which results from a reversible interconversion of S and Li_2S via polysulfide intermediaries [129, 130]. However, lithium polysulfide intermediates (Li_2S_x, $4 \leq \times \leq 8$) are soluble in ordinary liquid organic electrolytes during cycling, resulting in a polysulfide shuttle across the electrodes, polluting the system and decreasing the cell's stability and overall performance [131]. As a result, mitigating this shuttle process has remained a primary emphasis in the practical development of Li-S batteries. With the immobilization of polysulfides on the sulfur cathode side, functional separators or interlayers can efficiently control the shuttle effect [132]. Because their numerous hydroxyl groups may firmly bind sulfur-containing groups, biopolymers are thought to have the essential functionality to trap such polysulfide ions [133]. It has been demonstrated that chitosan effectively traps polysulfides in Li-S batteries by strong lithium polysulfide chemisorption, limiting sulfur loss into the electrolyte. Chen et al. coated a carbon membrane with chitosan to produce Li-S batteries with a capacity of 830 mAh g^{-1} at C/2 after 100 cycles, compared to 680 mAh g^{1} for a typical separator battery [134, 135]. The chitosan layer, which effectively traps and chemically immobilizes polysulfides in the cathode compartment, was credited with the improvement. Motivated by air-filtration principles, where filters must trap pollutants with minimal flow resistance, a gelatin-based polysulfide nanofilter was created by a protein-directed self-assembly approach [136]. Figure 2.10 shows the versatility of biopolymer in energy storage applications.

To eliminate polysulfide shuttling, a gel-like or solid electrolyte could be used instead of the traditional liquid electrolyte and separator pair. Although PEO and PVDF-HFP have been utilized to produce electrolytes that successfully prevent polysulfide dissolution, their weak ionic conductivities result in poor electrochemical performances [137, 138]. A sucrose-based SPE with a lithium ionic conductivity of 0.428 mS cm^{-1} and a Li^+ transference number of 0.76 was produced to improve the performance of PEO SPEs. After 257 cycles of reacting sucrose and borane-tetrahydrofuran complex, then adding PEO and LiTFSI salt, the resulting SPE has a capacity of 620 mAh g^{-1} at C/10.

Another study used lignocellulose fibers ranging from 50–300 μm obtained after wood treatment to produce GPEs. This was done using a solution casting method, followed by soaking the membrane in a solution of 1M LiTFSI in 1,2-dioxolane/dimethoxymethane (1:1 by volume) with 1 wt.% $LiNO_3$ for 2 hours [139, 140]. The porous nature of the GPE with its many cavities resulted in an ionic conductivity of 6.52 mS cm^{-1} and an electrochemical stability window of 5.3 V. Hydroxyl groups on lignocelluloses create hydrogen bonds with lithium salt anions and extremely electronegative polysulfides, resulting in a high Li+ transference number of 0.79 and reducing the shuttle effect by immobilizing polysulfides on the cathode side of the GPE. As a result, after 100 cycles, a Li-S cell with a capacity of 653 mAh g^{-1} at 20 mA g^{-1} was achieved.

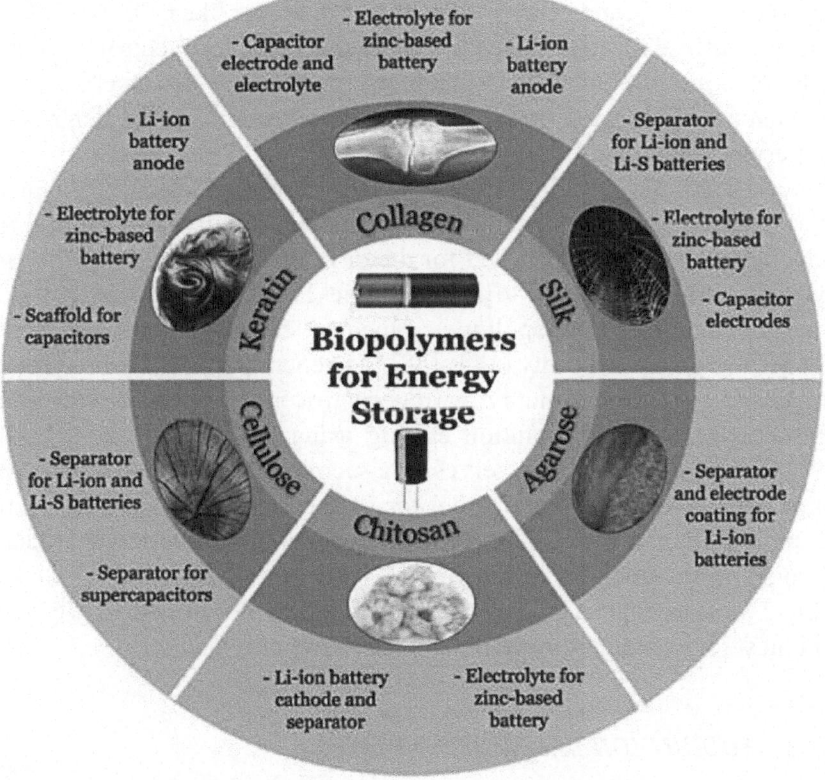

FIGURE 2.10 Diagram for biopolymers use in energy storage applications.

Source: Reprinted with permission from Ref. [215]. (Originally published on pixabay.com) https://creativecommons.org/licenses/by/4.0/

2.2.2.2 LITHIUM-OXYGEN (LI-O_2) BATTERIES

Aprotic lithium-oxygen batteries (Li-O_2) have piqued the interest of many since they have the potential to produce a far higher energy density than ordinary LIBs [141]. Separators based on polymers such as PVP or PVDFHF are not recommended for Li-O_2 batteries, as they react with superoxide/peroxide and reactive intermediates, resulting in undesirable side products. After Bruce and colleagues' fundamental work in 2006, Li-O_2 batteries are typically constituted of a Li metal anode [142], a glass microfiber membrane bathed in an ether-based Li^+ conducting electrolyte, and a porous cathode in contact with an oxygen environment.

Despite some early rapid progress, a number of significant challenges have stymied further growth. The leakage of volatile liquid electrolytes through the open cathode is a specific source of worry. Furthermore, if air is used, oxygen [143] (and other species found in the air like CO_2, N_2, H_2O, and so on) can diffuse through the cathode and the separator/electrolyte (oxygen can dissolve into the electrolyte) and reach the anode. This can cause passivating coatings to develop on the Li anode, resulting in a rapid deterioration of battery performance.

Because anionic biopolymers are capable of offering a significant number of coordination sites for formal cations, boosting ion dissociation and ionic conductivity, they are well suited for the development of Li-O_2 SPEs (solid polymer electrolytes) [144–146]. Many studies have demonstrated that adding extra functionalities to biopolymers, such as carboxymethyl groups, can boost their ionic conductivity [147, 148]. To generate a SPE, carboxymethyl-carrageenan and carboxymethyl-carrageenan were complexed with lithium nitrate ($LiNO_3$) through solution casting using a 1% acetic acid solution [149]. At the cathode, $LiNO_3$ serves as a secondary source of oxygen while also providing a nitrate ion channel for oxygen reduction. $LiNO_3$ interacts with the polymer hosts' COO and C-O-C groups, facilitating dissociation and boosting the quantity of ion routes [150]. A maximum t_{Li+} of 0.54 was obtained for κ-carrageen an derived SPEs (0.39 for PEO/LiPBSi) [151], with the highest ionic conductivities of about 5.85 mS cm^{-1} and 20% wt.% $LiNO_3$.

2.2.2.3 SODIUM-ION BATTERIES (NIBS)

Sodium ion batteries (NIBs) are seen as a sustainable and affordable option to lithium batteries. Because to the abundance of sodium resources (2.36

wt.% of the Earth's crust), their inexpensive cost, and a reasonable redox potential of 2.71 V versus standard hydrogen electrode (SHE), anode and cathode representation of sodium ion batteries can be seen in Figure 2.11 [152, 153]. The creation of superior NIB separators based on affordable renewable materials has consistently taken cues from the work done in the realm of LIBs.

Chen and co-workers used electrospun cellulose acetate (CA) nonwoven separators with a high specific surface area (87%) [154]. The intrinsic solubility of CA in common electrolytes was addressed by converting the surface acetyl groups into hydroxyl moieties using a 0.06 m NaOH solution, resulting in a separator with strong affinity towards EC/PC, EC/DMC, diglyme, and triglyme electrolytes (uptake of up to 518%). CA separators demonstrated tensile strength of 11 MPa and elongation at break of 9% with no shrinkage at 220?C. When evaluated in a $Na_3V_2(PO_4)_3$(NVP)/Na cell, an ultralong cycle life with a capacity retention of 93.78% after 10 000 cycles and a high rate capacity of 100.1 mAh g^{-1} at 10 C was achieved.

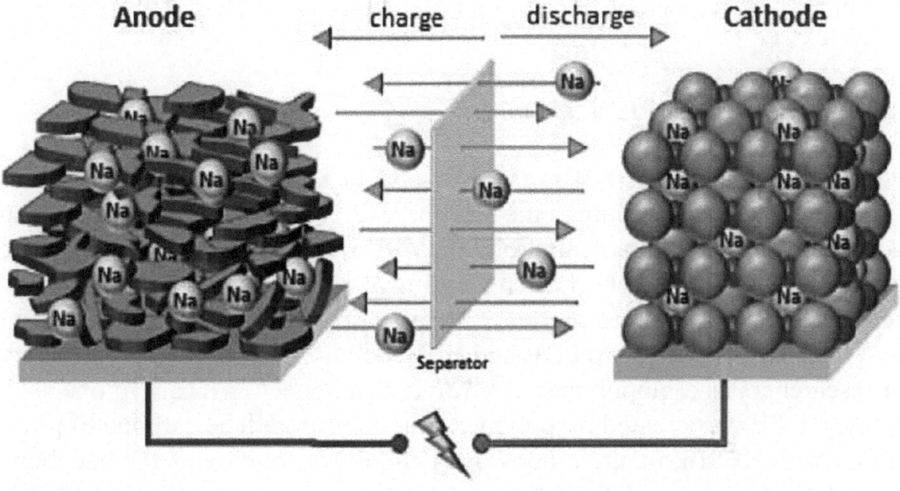

FIGURE 2.11 Sodium ion batteries (anode and cathode view).

Source: Reprinted with permission from Ref. [214] Copyright © 2019 by the authors. Licensee MDPI, Basel, Switzerland. http://creativecommons.org/licenses/by/4.0/

Most crucially, for Na cycling, a surprising reversible voltage response with square wave shape polarization profile was established, demonstrating very stable Na plating and stripping with Na dendrite development inhibition. The obtained improvements were attributed to a combination of efficient electrolyte infusion into the highly [155] porous structure, superior mechanical stability provided by crosslinking, and a smooth separator-electrode functionality offered by the inherent adhesive properties of the cellulose derivatives utilized. Zhang with his team employed a salt templating approach to create NIB separators with customizable pore size from prawn shell chitin nanofiber membranes [156]. In contrast to the hoover drying procedure, which produces dense structures, nanopore structures were weaved into chitin nanofiber membranes during the self-assembly process by introducing sodium dihydrogen citrate (SDCA) [157]. SDCA prevents nanofibers from tightly stacking during water evaporation, resulting in mesoporous films with a 252% electrolyte absorption (1 m $LiPF_6$ in EC/DEC).

The chitin separator demonstrated excellent mechanical properties, thermal stability up to 150?C, and a broad electrochemical stability window. Even though some papers exist on the production of GPEs and SPEs based on petroleum-derived polymers such as poly(vinylidene fluoride-co-hexafluoropropylene) (PVDF-HFP) [158], no reports on the development of GPEs or SPEs for NIBs made from natural biopolymers could be discovered.

2.2.2.4 POTASSIUM-ION BATTERIES (PIBS)

Potassium ion batteries (KIBs) are also appealing alternatives to LIBs due to the abundance of potassium in the biosphere (2.09 wt.% of Earth's crust) and potassium's low redox potential of 2.92 V vs SHE, which is extremely close to that of Li [132]. K^+ short Stokes radius of K+ allows for very fast ionic conductivities in bulk electrolytes, with three times greater ion diffusion than Li^+ [159]. In comparison to LIBs and NIBs, cellulose has piqued the curiosity of researchers as a biopolymer electrolyte component in KIBs. In one such study, a GPE was created by polymerizing poly(propylene carbonate) [160] (PPC) with KFSI utilizing a nonwoven cellulose structure as the backbone and tetrahydrofuran (THF) as the solvent.

The continuous and tenacious cellulose structure with interconnected pores can hold the solid electrolyte, resulting in an all-solid-state electrolyte that is homogeneous. At 18 wt.% salt content, the prepared SPE is stable up to 4.15 V (K/K^+) and has a maximal ionic conductivity of 0.0136 mS cm^{-1} following 40 cycles at 20 mA g^1. An all-solid-state 3,4,9,10-perylene-tetracarboxylicacid-dianhydride (PTCDA)/K battery with an all-solid-state 3,4,9,10-perylene-tetracarboxylicacid-dianhydride (PTCDA)/K electrolyte produced a capacity of 91.2 mAh g^1, suggesting efficient K metal stripping/plating.

2.2.3 SENSORS

High-quality, flexible, and extensible sensors that can properly perceive external signals have evolved into an important study subject in recent years, thanks to advancements in flexible electronics, materials science, and micro/nanoprocessing technology [161–164]. Natural biopolymers and chemically produced polymers with intrinsic biocompatibility have been widely used as active sensing layers, conductive electrodes, and substrates in a range of sensors [165, 166]. The incorporation of functional biopolymers by doping, surface modification, and chemical group design not only imbues the sensors with superior flexibility but also increases charge transport in the active sensing layer, hence boosting sensor sensitivity and detection limitations [167, 168]. This section focuses on chemical sensors, biosensors, and strain sensors that have performance improvements due to molecular alteration of biopolymers.

Biosensors have been extensively studied in the bioanalytical field, including biomarker identification [169, 170], drug release monitoring [171], and forensic analysis [172, 173]. Biological sensing could be performed via superior electrochemical activity by grafting or adding conducting polymers or molecules into biopolymers, which can supplement and increase the performance of biopolymers in biosensors.

Hui and co-workers [174] created unique composite nanofibers for DNA sensing by covalently grafting polyethylene glycol (PEG) polymer onto polyaniline (PANI) nanofibers, where PEG reduces nonspecific protein adsorption and hence increases specificity to cancer genes [107, 108]. Hyaluronic acid (HA), a glycosaminoglycan that is biodegradable, biocompatible, and non-immunogenic, has been intensively researched for biomedical uses.

Wei and co-workers [175] created a photo-electrochemical biosensor that uses the HA to bind CD44 receptors, allowing for high-sensitivity detection of CD44-related malignancies.

Silk is a representative natural biopolymer that has received a lot of interest in the field of flexible stretchable sensors [176, 177]. Liu and co-workers created a silk fibroin-based adhesive with micropillar structures on a polydimethylsiloxane (PDMS) substrate which demonstrated dependable and stabilized bonding force on skin surfaces under varying humidity conditions. The microstructure of fibroin adhesive is showing increasing promise for use in various epidermal electrical sensors [178].

Wang and his teammates created a very sensitive strain sensor based on carbonized silk nanofibers (CSNs), which outperformed other strain sensors in detecting human physiological signals. The nitrogen-doping carbon nanofiber network topology of CSNs is credited with the increased sensing capability [179, 180]. Figure 2.12 shows the fabrication process used for making the fibroin adhesive for the human skin of the wrist.

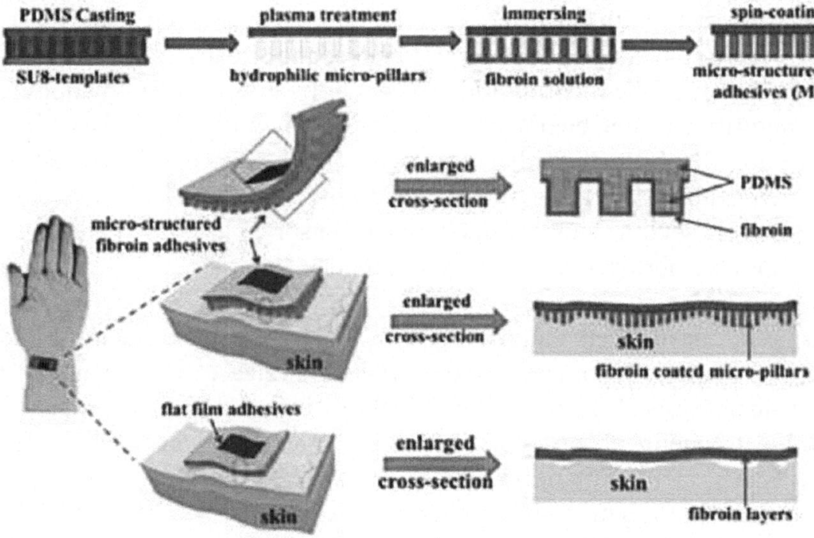

FIGURE 2.12 Fabrication process of micro-structured fibroin adhesive for the skin of wrist. *Source*: Reprinted with permission from Ref. [213]. https://creativecommons.org/licenses/by/4.0/

Strain sensors, particularly wearable strain sensors, have piqued the interest of researchers due to their potential application as an essential component in next-generation flexible and wearable electronic devices [181, 182]. High mechanical stretch ability has become one of the most important features for strain sensors in order to realize long-term monitoring of human health and activities [183]. The vast majority of biopolymers that are intrinsically brittle and have low chemical stability are rarely employed directly in flexible strain sensors. Nevertheless, annealing or doping with specific molecules can clearly improve biopolymer stretch factor and expand their application in strain sensors.

2.2.4 DISPLAYS

Flexible displays and lighting panels are now mass-produced, and several alternative flat panel display (FPD) technologies, such as active matrix liquid crystal displays (AM-LCDs), active matrix organic light-emitting displays (AM-OLEDs) [184, 185], have been put into the market. In addition, due to its inherent advantages of light weight, high brightness, low power consumption, comprehensive color area, fast response, and high flexibility, organic light-emitting diode (OLED) has been used as the preferred technology for flexible or conformal displays in the mobile display market [186]. With the demands of flexible and wearable technology for device biocompatibility and the advancement of bio-photonic research, it is critical to optimize OLED for biocompatibility and human affinity. Furthermore, biopolymers with broad accessibility, optical properties, and some biological activities, such as deoxyribonucleic acid (DNA) [187], chitosan, silk fibroin, cellulose, polyester, and hyaluronic acid, are interesting elements for further expanding display and lighting systems [188].

Natural biopolymers offer non-toxicity, great mechanical strength, strong biocompatibility, and biodegradability, making them suitable for use as substrates for flexible green electrical devices. Jin and co-workers [189] reported the development of a transparent paper based on self-assembled chitin nanofibers (CHnF) with strong hydrogen bonding. The OLED device is built on a 40 μm thick CHnF paper with stable flexibility, particularly in the bent condition. Furthermore, one of the major technical issues in the display business is the development of dependable and stable white light emitting diodes (WLED), while the use of low-cost coating is an effective technique for LED source conversion. Gotta and colleagues [190] used biocomposites of biological (crystalline nanocellulose and porcine gastric mucus) and organic (luminescent dye) components as coatings in blue LEDs to achieve white light emission.

Fujisaki and co-workers created unique transparent OTFT arrays based on CNF. Between solvated CNF particles, hydrogen bonding interactions exist, providing a stable three-dimensional network cross-linked structure. As a result, nanopaper constructed only of natural CNF has high thermal stability and chemical durability, allowing the created OTFT arrays to outperform devices made of standard glass. Wang and colleagues [191–195] demonstrated the production of top-gate OFETs on recyclable cellulose nanocrystal glycerol (CNC/glycerol) substrates, implying that CNF has evolved into an appealing flexible electronics material.

2.2.5 FABRICATION TECHNIQUES FOR PRINTING

2.2.5.1 SCREEN PRINTING

Screen printing is a printing technique that transfers ink to a substrate by squeezing high-viscosity inks through a stencil screen. With the fast evolution of printed electronics in recent years, screen printing has received widespread attention in the printing of photoelectric devices such as printed solar cells [196], organic light emitting diodes [197], organic field effect transistors [198], and radiofrequency tags, among others. Chouler and teammates [199] announced the first paper microbial fuel cell (pMFC) constructed by screen-printing biodegradable cellulose-based electrodes onto a single paper sheet, allowing for low-cost and straightforward device construction. More precisely, the screen-printing technique is also appropriate for fabricating biopolymer-based flexible electronic devices, providing excellent fabrication controllability while also protecting biopolymers from harm throughout the production process [200]. Ferri and co-workers [201] created a textile capacitive touchless sensor for hand gesture identification in which the electrodes are printed on textile substrates using screen printing technique, which is very efficient and preserves the textile's characteristics. As a result, the screen-printing technology is particularly promising for producing low-cost and environmentally friendly biopolymer-based electronics.

2.2.5.2 SPRAY PRINTING

At the moment, one of the most important methods of solution film formation is spin-coating. The spin-coating process, on the other hand, has a very high need for matching the surface properties of the rotatable material with the substrate surface. Furthermore, the low material usage ratio and uneven film limit its employment in low-cost and high-yield devices. Spray-coating has been widely used in flexible electronic devices to solve the disadvantages of the spin-coating approach, due to its great compatibility with biopolymers, high-efficiency production, and low-cost fabrication process [202].

Lee and co-workers [203] employed an ultrasonic nozzle to create organic poly(3,4-ethylenedioxythiophene):poly(styrenesulfonate) (PEDOT:PSS) electrodes with high reproducibility and uniformity for large-area and low-cost OFETs, demonstrating improved mobility reproducibility.

2.2.5.3 INKJET PRINTING

Inkjet printing works as a non-contact approach by gently dropping inks on a substrate to generate pre-designed patterned films, which have been widely utilized to build various devices such as transistors [204–206], solar cells, and sensors. In contrast to the previous two printing technologies, inkjet printing can provide high-resolution printing of electronically functional biopolymer inks on a variety of substrates, which holds promise for the construction of micrometer-scale devices with customizable patterns. Zimmermann and teammates [207], for example, used an inkjet printing process to create a silver and PEDOT: PSS electrode with biodegradability and biocompatibility for light-emitting electrochemical cells, demonstrating potential applications in life science and bioelectronics. Conti and co-workers [208] recently reported on the use of room-temperature inkjet-printed cellulose nanopapers (CNPs) as both the electrode and the substrate for innovative high-performance organic diodes, demonstrating improved rectification ratios. Inkjet printing is thus a very promising route for biopolymers to construct complex and low-cost electronic devices.

2.2.5.4 3D PRINTING

3D printing is an important advanced additive manufacturing method that can precisely construct complex 3D architectures without wasting materials for a variety of useful uses. Furthermore, 3D printing in biopolymer inks can be used without compromising bioactivity or importing harmful ingredients for biomedical and implantable electronics. As a result, the application of 3D printing to smart materials and biopolymers is becoming increasingly important in order to increase the use of flexible 3D electrical devices in tissue engineering [209], soft robots [210], and stretchable electronics [211]. Zhang and colleagues [212] employed 3D printing to create core-sheath structured fibers with CNTs as the conductive core and SF as the dielectric sheath for multifunctional energy management, E-textile, which combines the traditional benefits of textiles with integrated electric components while maintaining the textile's breathability and flexibility. As previously demonstrated, significant progress has been made in the printing approach of creating biopolymer-based electronics. When selecting the most appropriate techniques, the following issues are addressed based on biopolymer inks and manufacturing technologies: printable biopolymers with suitable rheological properties, such as viscosity, concentration, and plasticity of biopolymer inks, and the compatibility of printing technique with biopolymers to avoid loss of bioactive components and relative performances. Furthermore, the wetting and dewetting kinetic mechanisms of biopolymer inks on diverse substrates are not fully apparent, implying that additional research is required to achieve predictable and precise patterning of biopolymer inks. As a result, more emphasis should be placed on optimizing printing methods based on various biopolymer inks, which is crucial for realizing the application of biopolymers in electronic devices.

2.3 CONCLUSION

With growing environmental concerns, the sustainability and repeatability of energy, sensing, and other functional electronics have piqued the interest of many. This chapter extensively summarizes a range of biopolymers with distinct molecular architectures and specialized electronic properties in the application of diverse electronics. The corresponding properties of biopolymers can be adapted for various functional electronics, such as increased ion transport for energy electronics, altered transformation of charge carriers for sensors, and enhanced crystallization of biopolymers for flexible electronics, by molecular modification to biopolymers *via* chemical bonds or physical composite to regulate the molecular arrangement. Furthermore, in conjunction with the ease of dissolution of biopolymers, representative printing technologies are introduced to avoid the loss of bioactive components and thermal deformation of biopolymers, resulting in high performance and high yield of functional electronics on a large scale, at a low cost, and in an environmentally friendly manner. As a result, the combination of printing technology with advanced biopolymers via molecular modification is particularly promising for use in high-performance electronics on a broad scale while remaining environmentally friendly. Nonetheless, realizing the actual application of biopolymers in functional electronics remains a difficulty. To begin, the intrinsic constraints of biopolymers, such as thermal/optical instability, difficulties in maintaining bioactivity, and solubility in electrolytes, impede the development of high-performance, reproducible electronics. Second, the underlying process of ion transport in biopolymer molecules remains unknown, making it difficult to build appropriate and optimal biopolymers to boost electron and ion transport efficiency in electronics. Finally, the development and synthesis of innovative biopolymers with superior optical, electrical, magnetic, thermal, and mechanical properties for next-generation electronics such as flexible displays, wearable sensors, and other smart electronic systems is rarely studied. In summary, the incorporation of biopolymers into functional electronics is critical for minimizing electronic waste pollution and boosting their uses in green and sustainable energy, smart and new electronic systems. Although there are still challenges for biopolymers in practical applications, it is promising to adjust the properties of biopolymers through molecular engineering to propel the development of biopolymer-based electronics with high performance and stability for energy, human-machine interaction, biomedicine, and other applications.

KEYWORDS

- **electronic biopolymer**
- **energy devices**
- **human-machine interaction**
- **molecular engineering**
- **printing technologies**
- **sensors**
- **storage applications**

REFERENCES

1. Ezeoha, S. L., & Ezenwanne, J. N., (2013). Production of biodegradable plastic packaging film from cassava starch. *IOSR Journal of Engineering*, *3*(10), 14–20.
2. Ali, R. R., Rahman, W. W. A., Ibrahim, N. B., & Kasmani, R. M., (2013). Starch-based biofilms for green packaging. *Developments in Sustainable Chemical and Bioprocess Technology*, 347–354.
3. Thomas, S., Gopi, S., & Amalraj, A., (2020). *Biopolymers and Their Industrial Applications: From Plant, Animal, and Marine Sources, to Functional Products*. Elsevier.
4. Rao, S. S., & Rekha, P. D., (2022). Biopolymers in cosmetics, pharmaceutical, and biomedical applications. In: *Biopolymers: Recent Updates, Challenges and Opportunities* (pp. 223–244). Cham: Springer International Publishing.

5. Taherimehr, M., YousefniaPasha, H., Tabatabaeekoloor, R., & Pesaranhajiabbas, E., (2021). Trends and challenges of biopolymer-based nanocomposites in food packaging. *Comprehensive Reviews in Food Science and Food Safety*, *20*(6), 5321–5344.
6. Khalid, M. Y., & Arif, Z. U., (2022). Novel biopolymer-based sustainable composites for food packaging applications: A narrative review. *Food Packaging and Shelf Life*, *33*, 100892.
7. Singh, R., Gautam, S., Sharma, B., Jain, P., & Chauhan, K. D., (2021). Biopolymers and their classifications. In: *Biopolymers and Their Industrial Applications* (pp. 21–44). Elsevier.
8. Liu, Y., Ahmed, S., Sameen, D. E., Wang, Y., Lu, R., Dai, J., & Qin, W., (2021). A review of cellulose and its derivatives in biopolymer-based for food packaging application. *Trends in Food Science & Technology*, *112*, 532–546.
9. Chybczyńska, K., Markiewicz, E., Grząbka-Zasadzińska, A., & Borysiak, S., (2019). Dielectric, magnetic, and mechanical properties of composites consisting of biopolymer chitosan matrix and hybrid spinel/cellulose filler. *Ceramics International*, *45*(7), 9468–9476.
10. Ilyas, R. A., Sapuan, S. M., Norrrahim, M. N. F., Yasim-Anuar, T. A. T., Kadier, A., Kalil, M. S., & Asyraf, M. R. M., (2020). Nanocellulose/starch biopolymer nanocomposites: Processing, manufacturing, and applications. In: *Advanced Processing, Properties, and Applications of Starch and Other Bio-based Polymers* (pp. 65–88). Elsevier.
11. Ruiz, C. A. S., Baca, S. Z., Van, D. B. L. A., Van, D. B. C., Wijffels, R. H., & Eppink, M. H., (2020). Selective fractionation of free glucose and starch from microalgae using aqueous two-phase systems. *Algal Research*, *46*, 101801.
12. Apriyanto, A., Compart, J., & Fettke, J., (2022). A review of starch, a unique biopolymer–Structure, metabolism and in planta modifications. *Plant Science*, 111223.
13. Agarwal, S., (2021). Major factors affecting the characteristics of starch-based biopolymer films. *European Polymer Journal*, *160*, 110788.
14. Duan, Q., Chen, Y., Yu, L., & Xie, F., (2022). Chitosan–gelatin films: Plasticizers/nanofillers affect chain interactions and material properties in different ways. *Polymers*, *14*(18), 3797.
15. Ahmad, M. S., Ali, R. R., Majid, R. A., & Mohamad, Z., (2022). Properties enhancement of packaging materials based on gelatin. *Environmental Quality Management*.
16. Al-Nimry, S., Dayah, A. A., Hasan, I., & Daghmash, R., (2021). Cosmetic, biomedical and pharmaceutical applications of fish gelatin/hydrolysates. *Marine Drugs*, *19*(3), 145.
17. Negm, N. A., Hefni, H. H., Abd-Elaal, A. A., Badr, E. A., & Abou, K. M. T., (2020). Advancement on modification of chitosan biopolymer and its potential applications. *International Journal of Biological Macromolecules*, *152*, 681–702.
18. Detsi, A., Kavetsou, E., Kostopoulou, I., Pitterou, I., Pontillo, A. R. N., Tzani, A., & Zoumpoulakis, P., (2020). Nanosystems for the encapsulation of natural products: The case of chitosan biopolymer as a matrix. *Pharmaceutics*, *12*(7), 669.
19. Mohammadi, A. A., Dehghani, M. H., Mesdaghinia, A., Yaghmaian, K., & Es' haghi, Z., (2020). Adsorptive removal of endocrine disrupting compounds from aqueous solutions using magnetic multi-wall carbon nanotubes modified with chitosan biopolymer based on response surface methodology: Functionalization, kinetics, and isotherms studies. *International Journal of Biological Macromolecules*, *155*, 1019–1029.

20. Chen, B., Long, F., Chen, S., Cao, Y., & Pan, X., (2020). Magnetic chitosan biopolymer as a versatile adsorbent for simultaneous and synergistic removal of different sorts of dyestuffs from simulated wastewater. *Chemical Engineering Journal*, *385*, 123926.
21. Jamróz, E., Janik, M., Juszczak, L., Kruk, T., Kulawik, P., Szuwarzyński, M., & Khachatryan, K., (2021). Composite biopolymer films based on a polyelectrolyte complex of furcellaran and chitosan. *Carbohydrate Polymers*, *274*, 118627.
22. Ilyas, R. A., Zuhri, M. Y. M., Norrahim, M. N. F., Misenan, M. S. M., Jenol, M. A., Samsudin, S. A., & Omran, A. A. B., (2022). Natural fiber-reinforced polycaprolactone green and hybrid biocomposites for various advanced applications. *Polymers*, *14*(1), 182.
23. Prasad, A., & Kandasubramanian, B., (2019). Fused deposition processing polycaprolactone of composites for biomedical applications. *Polymer-Plastics Technology and Materials*, *58*(13), 1365–1398.
24. Abdelrazek, E. M., Hezma, A. M., El-Khodary, A., & Elzayat, A. M., (2016). Spectroscopic studies and thermal properties of PCL/PMMA biopolymer blend. *Egyptian Journal of Basic and Applied Sciences*, *3*(1), 10–15.
25. Wu, C. S., (2003). Physical properties and biodegradability of maleated-polycaprolactone/starch composite. *Polymer Degradation and Stability*, *80*(1), 127–134.
26. Salmoria, G. V., Klauss, P., Zepon, K. M., & Kanis, L. A., (2013). The effects of laser energy density and particle size in the selective laser sintering of polycaprolactone/progesterone specimens: Morphology and drug release. *The International Journal of Advanced Manufacturing Technology*, *66*, 1113–1118.
27. Jenkins, M. J., & Harrison, K. L., (2006). The effect of molecular weight on the crystallization kinetics of polycaprolactone. *Polymers for Advanced Technologies*, *17*(6), 474–478.
28. Tiptipakorn, S., Keungputpong, N., Phothiphiphit, S., & Rimdusit, S., (2015). Effects of polycaprolactone molecular weights on thermal and mechanical properties of polybenzoxazine. *Journal of Applied Polymer Science*, *132*(18).
29. Bartnikowski, M., Dargaville, T. R., Ivanovski, S., & Hutmacher, D. W., (2019). Degradation mechanisms of polycaprolactone in the context of chemistry, geometry and environment. *Progress in Polymer Science*, *96*, 1–20.
30. Nair, N. R., Sekhar, V. C., Nampoothiri, K. M., & Pandey, A., (2017). Biodegradation of biopolymers. In: *Current Developments in Biotechnology and Bioengineering* (pp. 739–755). Elsevier.
31. Kamoun, E. A., Kenawy, E. R. S., & Chen, X., (2017). A review on polymeric hydrogel membranes for wound dressing applications: PVA-based hydrogel dressings. *Journal of Advanced Research*, *8*(3), 217–233.
32. Abdullah, Z. W., Dong, Y., Davies, I. J., & Barbhuiya, S., (2017). PVA, PVA blends, and their nanocomposites for biodegradable packaging application. *Polymer-Plastics Technology and Engineering*, *56*(12), 1307–1344.
33. Nagarkar, R., & Patel, J., (2019). Polyvinyl alcohol: A comprehensive study. *Acta Scientific Pharmaceutical Sciences*, *3*(4), 34–44.
34. Kumar, A., & Sinha-Ray, S., (2018). A review on biopolymer-based fibers via electrospinning and solution blowing and their applications. *Fibers*, *6*(3), 45.
35. Rafique, A., Zia, K. M., Zuber, M., Tabasum, S., & Rehman, S., (2016). Chitosan functionalized poly (vinyl alcohol) for prospects biomedical and industrial applications: A review. *International Journal of Biological Macromolecules*, *87*, 141–154.

36. Deshmukh, K., Ahamed, M. B., Sadasivuni, K. K., Ponnamma, D., Al Maadeed, M. A. A., Deshmukh, R. R., & Chidambaram, K., (2017). Fumed SiO_2 nanoparticle reinforced biopolymer blend nanocomposites with high dielectric constant and low dielectric loss for flexible organic electronics. *Journal of Applied Polymer Science*, *134*(5).
37. Maria, T. M., De Carvalho, R. A., Sobral, P. J., Habitante, A. M. B., & Solorza-Feria, J., (2008). The effect of the degree of hydrolysis of the PVA and the plasticizer concentration on the color, opacity, and thermal and mechanical properties of films based on PVA and gelatin blends. *Journal of Food Engineering*, *87*(2), 191–199.
38. Abdullah, Z. W., Dong, Y., Han, N., & Liu, S., (2019). Water and gas barrier properties of polyvinyl alcohol (PVA)/starch (ST)/glycerol (GL)/halloysite nanotube (HNT) bionanocomposite films: Experimental characterization and modelling approach. *Composites Part B: Engineering*, *174*, 107033.
39. Khwaldia, K., Arab-Tehrany, E., & Desobry, S., (2010). Biopolymer coatings on paper packaging materials. *Comprehensive Reviews in Food Science and Food Safety*, *9*(1), 82–91.
40. Vineeth, S. K., Gadhave, R. V., & Gadekar, P. T., (2022). Polyvinyl alcohol–cellulose blend wood adhesive modified by citric acid and its effect on physical, thermal, mechanical and performance properties. *Polymer Bulletin*, 1–18.
41. Hao, N. H., Tung, N. H., Van, T. M., & Nhan, N. T. T., (2015). Relationship between the structure of polymer network using glutaraldehyde crosslinking agent and physico-mechanical properties of biopolymer membrane based on polyvinyl alcohol (PVA) modified with cassava starch. *Vietnam Journal of Chemistry*, *53*(2), 188–193.
42. Inda, N. I., Pato, V., Maraeka, G., Sosidi, H., & Satrimafitrah, P., (2022). Preparation of pellet biopolymer polyvinyl alcohol/alginate/gluataraldehyde impregnated with banana peel activated carbon (BPAC). *Gravitasi*, *21*(2), 65–68.
43. Rault, J., Gref, R., Ping, Z. H., Nguyen, Q. T., & Néel, J., (1995). Glass transition temperature regulation effect in a poly (vinyl alcohol)—Water system. *Polymer*, *36*(8), 1655–1661.
44. Zhang, H., Niu, C., Zhang, Y., Wang, X., & Yang, B., (2020). A mechanically strong polyvinyl alcohol/poly (2-(N, N′-dimethyl amino) ethyl methacrylate)-poly (acrylic acid) hydrogel with pH-responsiveness. *Colloid and Polymer Science*, *298*, 619–628.
45. Taylor, R. D., Jewsbury, P. J., & Essex, J. W., (2002). A review of protein-small molecule docking methods. *Journal of Computer-Aided Molecular Design*, *16*, 151–166.
46. Wang, K., Zhou, C., Hong, Y., & Zhang, X., (2012). A review of protein adsorption on bioceramics. *Interface Focus*, *2*(3), 259–277.
47. Al-Lazikani, B., Jung, J., Xiang, Z., & Honig, B., (2001). Protein structure prediction. *Current Opinion in Chemical Biology*, *5*(1), 51–56.
48. Wang, B., Yang, W., McKittrick, J., & Meyers, M. A., (2016). Keratin: Structure, mechanical properties, occurrence in biological organisms, and efforts at bioinspiration. *Progress in Materials Science*, *76*, 229–318.
49. Rajabinejad, H., Bucişcanu, I. I., & Maier, S. S., (2016). Practical ways of extracting keratin from keratinous wastes and by-products: A review. *Environmental Engineering & Management Journal (EEMJ)*, *15*(5).
50. Barsotti, M. C., Felice, F., Balbarini, A., & Di Stefano, R., (2011). Fibrin as a scaffold for cardiac tissue engineering. *Biotechnology and Applied Biochemistry*, *58*(5), 301–310.
51. Reis, C. H. B., Buchaim, D. V., Ortiz, A. D. C., Fideles, S. O. M., Dias, J. A., Miglino, M. A., & Buchaim, R. L., (2022). Application of fibrin associated with photobiomodulation

as a promising strategy to improve regeneration in tissue engineering: A systematic review. *Polymers*, *14*(15), 3150.
52. Chrobak, M. O., Hansen, K. J., Gershlak, J. R., Vratsanos, M., Kanellias, M., Gaudette, G. R., & Pins, G. D., (2017). Design of a fibrin microthread-based composite layer for use in a cardiac patch. *ACS Biomaterials Science & Engineering*, *3*(7), 1394–1403.
53. Shishir, M. R. I., Xie, L., Sun, C., Zheng, X., & Chen, W., (2018). Advances in micro and nano-encapsulation of bioactive compounds using biopolymer and lipid-based transporters. *Trends in Food Science & Technology*, *78*, 34–60.
54. Basim, P., Gorityala, S., & Kurakula, M., (2021). Advances in functionalized hybrid biopolymer augmented lipid-based systems: A spotlight on their role in design of gastro retentive delivery systems. *Archives of Gastroenterology Research*, *2*(1), 35–47.
55. Singh, R., Polu, A. R., Bhattacharya, B., Rhee, H. W., Varlikli, C., & Singh, P. K., (2016). Perspectives for solid biopolymer electrolytes in dye sensitized solar cell and battery application. *Renewable and Sustainable Energy Reviews*, *65*, 1098–1117.
56. Brindha, K., Mohanraj, S., Rajaguru, P., & Pugalenthi, V., (2023). Simultaneous production of renewable biohydrogen, biobutanol and biopolymer from phytogenic CoNPs-assisted Clostridial fermentation for sustainable energy and environment. *Science of The Total Environment*, *859*, 160002.
57. Devadas, V. V., Khoo, K. S., Chia, W. Y., Chew, K. W., Munawaroh, H. S. H., Lam, M. K., & Show, P. L., (2021). Algae biopolymer towards sustainable circular economy. *Bioresource Technology*, *325*, 124702.
58. Posen, I. D., Jaramillo, P., Landis, A. E., & Griffin, W. M., (2017). Greenhouse gas mitigation for US plastics production: Energy first, feedstocks later. *Environmental Research Letters*, *12*(3), 034024.
59. Zhang, Y., Schnepp, Z., Cao, J., Ouyang, S., Li, Y., Ye, J., & Liu, S., (2013). Biopolymer-activated graphitic carbon nitride towards a sustainable photocathode material. *Scientific Reports*, *3*(1), 2163.
60. Rasheed, T., Anwar, M. T., Naveed, A., & Ali, A., (2022). Biopolymer based materials as alternative greener binders for sustainable electrochemical energy storage applications. *Chemistry Select*, *7*(39), e202203202.
61. Suvarnna, K., Kirubavathy, S. J., Selvasekarapandian, S., Krishna, M. V., & Ramaswamy, M., (2022). Corn silk extract–based solid-state biopolymer electrolyte and its application to electrochemical storage devices. *Ionics*, *28*(4), 1767–1782.
62. Nagaraju, D. H., Rebis, T., Gabrielsson, R., Elfwing, A., Milczarek, G., & Inganäs, O., (2014). Charge storage capacity of renewable biopolymer/conjugated polymer interpenetrating networks enhanced by electroactive dopants. *Advanced Energy Materials*, *4*(1), 1300443.
63. Aziz, S. B., Hamsan, M. H., Nofal, M. M., San, S., Abdulwahid, R. T., Raza, S. S., & Al-Zangana, S., (2020). From cellulose, shrimp and crab shells to energy storage EDLC cells: The study of structural and electrochemical properties of proton conducting chitosan-based biopolymer blend electrolytes. *Polymers*, *12*(7), 1526.
64. Zheng, Q., Zhang, H., Shi, B., Xue, X., Liu, Z., Jin, Y., & Wang, Z. L., (2016). *In vivo* self-powered wireless cardiac monitoring via implantable triboelectric nanogenerator. *ACS Nano*, *10*(7), 6510–6518.
65. Zheng, Q., Shi, B., Fan, F., Wang, X., Yan, L., Yuan, W., & Wang, Z. L., (2014). *In vivo* powering of pacemaker by breathing-driven implanted triboelectric nanogenerator. *Advanced Materials*, *26*(33), 5851–5856.

66. Zi, Y., Wang, J., Wang, S., Li, S., Wen, Z., Guo, H., & Wang, Z. L., (2016). Effective energy storage from a triboelectric nanogenerator. *Nature Communications*, *7*(1), 10987.
67. Kim, H. J., Kim, J. H., Jun, K. W., Kim, J. H., Seung, W. C., Kwon, O. H., & Oh, I. K., (2016). Silk nanofiber-networked bio-triboelectric generator: Silk bio-TEG. *Advanced Energy Materials*, *6*(8), 1502329.
68. Kim, H. J., Yim, E. C., Kim, J. H., Kim, S. J., Park, J. Y., & Oh, I. K., (2017). Bacterial nano-cellulose triboelectric nanogenerator. *Nano Energy*, *33*, 130–137.
69. An, S., Sankaran, A., & Yarin, A. L., (2018). Natural biopolymer-based triboelectric nanogenerators via fast, facile, scalable solution blowing. *ACS Applied Materials & Interfaces*, *10*(43), 37749–37759.
70. Torres, F. G., & De-la-Torre, G. E., (2021). Polysaccharide-based triboelectric nanogenerators: A review. *Carbohydrate Polymers*, *251*, 117055.
71. Ma, C., Gao, S., Gao, X., Wu, M., Wang, R., Wang, Y., & Wu, W., (2019). Chitosan biopolymer-derived self-powered triboelectric sensor with optimized performance through molecular surface engineering and data-driven learning. *InfoMat*, *1*(1), 116–125.
72. Bai, L., Li, Q., Yang, Y., Ling, S., Yu, H., Liu, S., & Chen, W., (2021). Biopolymer nanofibers for nanogenerator development. *Research*, *2021*.
73. Charoonsuk, T., Pongampai, S., Pakawanit, P., & Vittayakorn, N., (2021). Achieving a highly efficient chitosan-based triboelectric nanogenerator via adding organic proteins: Influence of morphology and molecular structure. *Nano Energy*, *89*, 106430.
74. O'regan, B., & Grätzel, M., (1991). A low-cost, high-efficiency solar cell based on dye-sensitized colloidal TiO_2 films. *Nature*, *353*(6346), 737–740.
75. Ghann, W., Kang, H., Sheikh, T., Yadav, S., Chavez-Gil, T., Nesbitt, F., & Uddin, J., (2017). Fabrication, optimization and characterization of natural dye sensitized solar cell. *Scientific Reports*, *7*(1), 41470.
76. Rajendran, S., Naushad, M., Raju, K., & Boukherroub, R., (2019). *Emerging Nanostructured Materials for Energy and Environmental Science*. Springer International Publishing.
77. Kusumawati, Y., Hutama, A. S., Wellia, D. V., & Subagyo, R., (2021). Natural resources for dye-sensitized solar cells. *Heliyon*, *7*(12), e08436.
78. Meier, H., Huang, Z. S., & Cao, D., (2017). Double D–π–A branched dyes–a new class of metal-free organic dyes for efficient dye-sensitized solar cells. *Journal of Materials Chemistry C*, *5*(38), 9828–9837.
79. Musazade, E., Voloshin, R., Brady, N., Mondal, J., Atashova, S., Zharmukhamedov, S. K., & Allakhverdiev, S. I., (2018). Biohybrid solar cells: Fundamentals, progress, and challenges. *Journal of Photochemistry and Photobiology C: Photochemistry Reviews*, *35*, 134–156.
80. Jang, H. S., Yun, J. M., Kim, D. Y., Park, D. W., Na, S. I., & Kim, S. S., (2012). Moderately reduced graphene oxide as transparent counter electrodes for dye-sensitized solar cells. *Electrochimica Acta*, *81*, 301–307.
81. Zahedi, R., Moosavian, S. F., & Aslani, A., (2022). Environmental and damage assessment of transparent solar cells compared with first and second generations using the LCA approach. *Energy Science & Engineering, 10*(12), 4640–4661.
82. Nagaraj, P., Sasidharan, A., David, V., & Sambandam, A., (2017). Effect of cross-linking on the performances of starch-based biopolymer as gel electrolyte for dye-sensitized solar cell applications. *Polymers*, *9*(12), 667.

83. Karthick, S. N., Hemalatha, K. V., Balasingam, S. K., Manik Clinton, F., Akshaya, S., & Kim, H. J., (2019). Dye-sensitized solar cells: history, components, configuration, and working principle. *Interfacial Engineering in Functional Materials for Dye-Sensitized Solar Cells*, 1–16.
84. Selvanathan, V., Ruslan, M. H., Alkahtani, A. A. N., Amin, N., Sopian, K., Muhammad, G., & Akhtaruzzaman, M., (2021). Organosoluble, esterified starch as quasi-solid biopolymer electrolyte in dye-sensitized solar cell. *Journal of Materials Research and Technology*, *12*, 1638–1648.
85. Cao, L., Yang, M., Wu, D., Lyu, F., Sun, Z., Zhong, X., & Lu, Z., (2017). Biopolymer-chitosan based supramolecular hydrogels as solid-state electrolytes for electrochemical energy storage. *Chemical Communications*, *53*(10), 1615–1618.
86. Shukur, M. F., Yusof, Y. M., Zawawi, S. M. M., Illias, H. A., & Kadir, M. F. Z., (2013). Conductivity and transport studies of plasticized chitosan-based proton conducting biopolymer electrolytes. *Physica Scripta*, *2013*(T157), 014050.
87. Ikram, S., Ahmed, S., Wazed, A. S., & Agarwal, H., (2017). Chitosan-based polymer electrolyte membranes for fuel cell applications. *Organic-Inorganic Composite Polymer Electrolyte Membranes: Preparation, Properties, and Fuel Cell Applications*, 381–398.
88. Fadzallah, I. A., Majid, S. R., Careem, M. A., & Arof, A. K., (2014). A study on ionic interactions in chitosan–oxalic acid polymer electrolyte membranes. *Journal of Membrane Science*, *463*, 65–72.
89. Aziz, S. B., Asnawi, A. S., Kadir, M. F. Z., Alshehri, S. M., Ahamad, T., Yusof, Y. M., & Hadi, J. M., (2021). Structural, electrical and electrochemical properties of glycerolized biopolymers based on chitosan (Cs): Methylcellulose (mc) for energy storage application. *Polymers*, *13*(8), 1183.
90. Jayanthi, S., & Ulaganathan, M., (2021). Biopolymer electrolytes for energy storage applications. In: *Polymer Electrolytes for Energy Storage Devices* (pp. 255–275). CRC Press.
91. Shukur, M. F., Ithnin, R., Illias, H. A., & Kadir, M. F. Z., (2013). Proton conducting polymer electrolyte based on plasticized chitosan–PEO blend and application in electrochemical devices. *Optical Materials*, *35*(10), 1834–1841.
92. Abdulkareem, S. S., (2021). Structural, morphological and electrical properties of chitosan/methylcellulose blend polymer doped with different concentrations of NH_4NO_3. *Materials Research Express*, *8*(8), 086301.
93. Rani, M. S. A., Rudhziah, S., Ahmad, A., & Mohamed, N. S., (2014). Biopolymer electrolyte based on derivatives of cellulose from kenaf bast fiber. *Polymers*, *6*(9), 2371–2385.
94. Monisha, S., Selvasekarapandian, S., Mathavan, T., Milton, F. B. A., Manoharan, S., & Karthikeyan, S., (2016). Preparation and characterization of biopolymer electrolyte based on cellulose acetate for potential applications in energy storage devices. *Journal of Materials Science: Materials in Electronics*, *27*, 9314–9324.
95. Salleh, N. S., Aziz, S. B., Aspanut, Z., & Kadir, M. F. Z., (2016). Electrical impedance and conduction mechanism analysis of biopolymer electrolytes based on methyl cellulose doped with ammonium iodide. *Ionics*, *22*, 2157–2167.
96. Samsudin, A. S., Kuan, E. C. H., & Isa, M. I. N., (2011). Investigation of the potential of proton-conducting biopolymer electrolytes based methyl cellulose-glycolic acid. *International Journal of Polymer Analysis and Characterization*, *16*(7), 477–485.

97. Rudhziah, S., Ahmad, A., Ahmad, I., & Mohamed, N. S., (2015). Biopolymer electrolytes based on blend of kappa-carrageenan and cellulose derivatives for potential application in dye sensitized solar cell. *Electrochimica Acta, 175*, 162–168.
98. Mahalakshmi, M., Selvanayagam, S., Selvasekarapandian, S., Moniha, V., Manjuladevi, R., & Sangeetha, P., (2019). Characterization of biopolymer electrolytes based on cellulose acetate with magnesium perchlorate ($Mg(ClO_4)_2$) for energy storage devices. *Journal of Science: Advanced Materials and Devices*, *4*(2), 276–284.
99. Shetty, S. K., Ismayil, & Shetty, G., (2020). Enhancement of electrical and optical properties of sodium bromide doped carboxymethyl cellulose biopolymer electrolyte films. *Journal of Macromolecular Science, Part B*, *59*(4), 235–247.
100. Nunes, S. C., Pereira, R. F., Sousa, N., Silva, M. M., Almeida, P., Figueiredo, F. M., & de Zea Bermudez, V., (2017). Eco-Friendly Red Seaweed-Derived Electrolytes for Electrochemical Devices. *Advanced Sustainable Systems*, *1*(9), 1700070.
101. Singh, R., Jadhav, N. A., Majumder, S., Bhattacharya, B., & Singh, P. K., (2013). Novel biopolymer gel electrolyte for dye-sensitized solar cell application. *Carbohydrate Polymers*, *91*(2), 682–685.
102. Selvalakshmi, S., Vijaya, N., Selvasekarapandian, S., & Premalatha, M., (2017). Biopolymer agar-agar doped with NH_4SCN as solid polymer electrolyte for electrochemical cell application. *Journal of Applied Polymer Science*, *134*(15).
103. Yang, Y., Cui, J., Yi, P., Zheng, X., Guo, X., & Wang, W., (2014). Effects of nanoparticle additives on the properties of agarose polymer electrolytes. *Journal of Power Sources*, *248*, 988–993.
104. Yang, Y., Hu, H., Zhou, C. H., Xu, S., Sebo, B., & Zhao, X. Z., (2011). Novel agarose polymer electrolyte for quasi-solid-state dye-sensitized solar cell. *Journal of Power Sources*, *196*(4), 2410–2415.
105. O'Regan, B., Schwartz, D. T., Zakeeruddin, S. M., & Grätzel, M., (2000). Electrodeposited nanocomposite n–p heterojunctions for solid-state dye-sensitized photovoltaics. *Advanced Materials*, *12*(17), 1263–1267.
106. Singh, R., Bhattacharya, B., Tomar, S. K., Singh, V., & Singh, P. K., (2017). Electrical, optical and electrophotochemical studies on agarose-based biopolymer electrolyte towards dye sensitized solar cell application. *Measurement*, *102*, 214–219.
107. Hsu, H. L., Tien, C. F., Yang, Y. T., & Leu, J., (2013). Dye-sensitized solar cells based on agarose gel electrolytes using allylimidazolium iodides and environmentally benign solvents. *Electrochimica Acta*, *91*, 208–213.
108. Guo, X., Yi, P., Yang, Y., Cui, J., Xiao, S., & Wang, W., (2013). Effects of surfactants on agarose-based magnetic polymer electrolyte for dye-sensitized solar cells. *Electrochimica Acta*, *90*, 524–529.
109. Ab Rahman, N. F., Shyuan, L. K., Mohamad, A. B., & Kadhum, A. A. H., (2013). Review on biopolymer membranes for fuel cell applications. *Applied Mechanics and Materials*, *291*, 614–617.
110. Yogananda, K. C., Ramasamy, E., Kumar, S., Kumar, S. V., Rani, M. N., & Rangappa, D., (2017). Novel rice starch based aqueous gel electrolyte for dye sensitized solar cell application. *Materials Today: Proceedings*, *4*(11), 12238–12244.
111. Ornaghi, Jr. H. L., Kerche, E. F., Neves, R. M., Monticeli, F. M., & Agnol, L. D., (2022). A systematic review of new trends in ionic liquids applied to electrolytes on polysaccharides. *Polysaccharides*, *3*(3), 502–514.

112. Khanmirzaei, M. H., & Ramesh, S., (2014). Nanocomposite polymer electrolyte based on rice starch/ionic liquid/TiO_2 nanoparticles for solar cell application. *Measurement*, *58*, 68–72.
113. Khanmirzaei, M. H., Ramesh, S., & Ramesh, K., (2015). Polymer electrolyte-based dye-sensitized solar cell with rice starch and 1-methyl-3-propylimidazolium iodide ionic liquid. *Materials & Design*, *85*, 833–837.
114. Muhammad, F. H., & Winie, T., (2020). Influence of 1-methyl-3-propylimidazolium iodide ionic liquid on the performance of dye-sensitized solar cell using hexanoyl chitosan/poly (vinyl chloride) based polymer electrolyte. *Optik*, *208*, 164558.
115. Lobregas, M. O. S., & Camacho, D. H., (2019). Gel polymer electrolyte system based on starch grafted with ionic liquid: Synthesis, characterization and its application in dye-sensitized solar cell. *Electrochimica Acta*, *298*, 219–228.
116. Min, H. J., Jung, J. H., Kang, M., & Kim, J. H., (2019). Synthesis of starch-g-PAN polymer electrolyte membrane and its application to flexible solid supercapacitors. *Membrane Journal*, *29*(3), 164–172.
117. D'Anna, F., Grilli, M. L., Petrucci, R., & Feroci, M., (2020). WO_3 and ionic liquids: A synergic pair for pollutant gas sensing and desulfurization. *Metals*, *10*(4), 475.
118. Karbowiak, T., Hervet, H., Léger, L., Champion, D., Debeaufort, F., & Voilley, A., (2006). Effect of plasticizers (water and glycerol) on the diffusion of a small molecule in iota-carrageenan biopolymer films for edible coating application. *Biomacromolecules*, *7*(6), 2011–2019.
119. Qamar, S. A., Junaid, M., Riasat, A., Jahangeer, M., Bilal, M., & Mu, B. Z., (2022). Carrageenan-based hybrids with biopolymers and nano-structured materials for biomimetic applications. *Starch-Stärke*, 2200018.
120. Chitra, R., Sathya, P., Selvasekarapandian, S., Monisha, S., Moniha, V., & Meyvel, S., (2019). Synthesis and characterization of iota-carrageenan solid biopolymer electrolytes for electrochemical applications. *Ionics*, *25*(5), 2147–2157.
121. Rayung, M., Aung, M. M., Azhar, S. C., Abdullah, L. C., Su'ait, M. S., Ahmad, A., & Jamil, S. N. A. M., (2020). Bio-based polymer electrolytes for electrochemical devices: Insight into the ionic conductivity performance. *Materials*, *13*(4), 838.
122. Zainuddin, N., & Samsudin, A. S., (2018). Investigation on the effect of NH_4Br at transport properties in k-carrageenan based biopolymer electrolytes via structural and electrical analysis. *Materials Today Communications*, *14*, 199–209.
123. Priya, S. S., Karthika, M., Selvasekarapandian, S., & Manjuladevi, R., (2018). Preparation and characterization of polymer electrolyte based on biopolymer I-carrageenan with magnesium nitrate. *Solid State Ionics*, *327*, 136–149.
124. Shamsudin, I. J., Ahmad, A., Hassan, N. H., & Kaddami, H., (2016). Biopolymer electrolytes based on carboxymethyl κ-carrageenan and imidazolium ionic liquid. *Ionics*, *22*, 841–851.
125. Ghani, N. A. A., Anuar, F. H., Ahmad, A., Mobarak, N. N., Shamsudin, I. J., Dzulkipli, M. Z., & Hassan, N. H., (2020). Incorporating 1-butyl-3-methylimidazolium chloride ionic liquid into iota carrageenan solid biopolymer electrolyte for electrochemical devices application. *Sains Malaysiana*, *49*(2), 305–313.
126. Nandhinilakshmi, M., Vanitha, D., Nallamuthu, N., Sundaramahalingam, K., & Saranya, P., (2022). Investigation on conductivity and optical properties for blend electrolytes based on iota-carrageenan and acacia gum with ethylene glycol. *Journal of Materials Science: Materials in Electronics*, *33*(26), 21172–21188.

127. Mobarak, N. N., Jumaah, F. N., Ghani, M. A., Abdullah, M. P., & Ahmad, A., (2015). Carboxymethyl carrageenan-based biopolymer electrolytes. *Electrochimica Acta*, *175*, 224–231.
128. Hasan, M. M., Islam, M. D., & Rashid, T. U., (2020). Biopolymer-based electrolytes for dye-sensitized solar cells: A critical review. *Energy & Fuels*, *34*(12), 15634–15671.
129. Shaibani, M., Mirshekarloo, M. S., Singh, R., Easton, C. D., Cooray, M. D., Eshraghi, N., & Majumder, M., (2020). Expansion-tolerant architectures for stable cycling of ultrahigh-loading sulfur cathodes in lithium-sulfur batteries. *Science Advances*, *6*(1), eaay2757.
130. He, J., & Manthiram, A., (2019). A review on the status and challenges of electrocatalysts in lithium-sulfur batteries. *Energy Storage Mater.*, *20*, 55–70.
131. Manthiram, A., Fu, Y., & Su, Y. S., (2013). Challenges and prospects of lithium–sulfur batteries. *Accounts of Chemical Research*, *46*(5), 1125–1134.
132. Manthiram, A., Chung, S. H., & Zu, C., (2015). Lithium–sulfur batteries: Progress and prospects. *Advanced Materials*, *27*(12), 1980–2006.
133. Zhang, T. W., Tian, T., Shen, B., Song, Y. H., & Yao, H. B., (2019). Recent advances on biopolymer fiber-based membranes for lithium-ion battery separators. *Composites Communications*, *14*, 7–14.
134. Winie, T., & Arof, A. K., (2016). Biopolymer electrolytes for energy devices. *Nanostruct. Polym. Membr. Appl.*, *2*, 311–356.
135. Kadir, M. F. Z., (2021). Non-faradaic-based supercapacitor fabricated with fish skin gelatin biopolymer electrolyte. *Ionics*, *27*, 2219–2229.
136. Li, F., Du, P., Chen, W., & Zhang, S., (2007). Preparation of silica-supported porous sorbent for heavy metal ions removal in wastewater treatment by organic–inorganic hybridization combined with sucrose and polyethylene glycol imprinting. *Analytica Chimica Acta*, *585*(2), 211–218.
137. Ruiz-Hitzky, E., Darder, M., Fernandes, F. M., Zatile, E., Palomares, F. J., & Aranda, P., (2011). Supported Graphene from Natural Resources: Easy Preparation and Applications. *Advanced Materials*, *23*(44), 5250–5255.
138. Musa, M. T., Shaari, N., Kamarudin, S. K., & Wong, W. Y., (2022). Recent biopolymers used for membrane fuel cells: Characterization analysis perspectives. *International Journal of Energy Research*, *46*(12), 16178–16207.
139. Song, A., Huang, Y., Zhong, X., Cao, H., Liu, B., Lin, Y., & Li, X., (2018). Novel lignocellulose-based gel polymer electrolyte with higher comprehensive performances for rechargeable lithium–sulfur battery. *Journal of Membrane Science*, *556*, 203–213.
140. Dalwadi, S., Goel, A., Kapetanakis, C., Salas-De, La. C. D., & Hu, X., (2023). The integration of biopolymer-based materials for energy storage applications: A review. *International Journal of Molecular Sciences*, *24*(4), 3975.
141. Sun, Y., (2013). Lithium ion conducting membranes for lithium-air batteries. *Nano Energy*, *2*(5), 801–816.
142. Kichambare, P., Rodrigues, S., & Kumar, J., (2012). Mesoporous nitrogen-doped carbon-glass ceramic cathodes for solid-state lithium–oxygen batteries. *ACS Applied Materials & Interfaces*, *4*(1), 49–52.
143. Leng, L., Zeng, X., Chen, P., Shu, T., Song, H., Fu, Z., & Liao, S., (2015). A novel stability-enhanced lithium-oxygen battery with cellulose-based composite polymer gel as the electrolyte. *Electrochimica Acta*, *176*, 1108–1115.

144. Mobarak, N. N., Ramli, N., Ahmad, A., & Rahman, M. Y. A., (2012). Chemical interaction and conductivity of carboxymethyl κ-carrageenan based green polymer electrolyte. *Solid State Ionics*, *224*, 51–57.
145. Jumaah, F. N., Mobarak, N. N., Ahmad, A., Ghani, M. A., & Rahman, M. Y. A., (2015). Derivative of iota-carrageenan as solid polymer electrolyte. *Ionics*, *21*, 1311–1320.
146. Bella, F., Mobarak, N. N., Jumaah, F. N., & Ahmad, A., (2015). From seaweeds to biopolymeric electrolytes for third generation solar cells: An intriguing approach. *Electrochimica Acta*, *151*, 306–311.
147. Mobarak, N. N., Ahmad, A., Abdullah, M. P., Ramli, N., & Rahman, M. Y. A., (2013). Conductivity enhancement via chemical modification of chitosan based green polymer electrolyte. *Electrochimica Acta*, *92*, 161–167.
148. Lu, N., Ho, Y. M., Fan, C. W., Wang, F. M., & Lee, J. T., (2007). A simple method for synthesizing polymeric lithium salts exhibiting relatively high cationic transference number in solid polymer electrolytes. *Solid State Ionics*, *178*(5, 6), 347–353.
149. Yabuuchi, N., Kubota, K., Dahbi, M., & Komaba, S., (2014). Research development on sodium-ion batteries. *Chemical Reviews*, *114*(23), 11636–11682.
150. Chen, W., Zhang, L., Liu, C., Feng, X., Zhang, J., Guan, L., & Cui, S., (2018). Electrospun flexible cellulose acetate-based separators for sodium-ion batteries with ultralong cycle stability and excellent wettability: The role of interface chemical groups. *ACS Applied Materials & Interfaces*, *10*(28), 23883–23890.
151. Yang, D., He, L., Liu, Y., Yan, W., Liang, S., Zhu, Y., & Wu, Y., (2019). An acetylene black modified gel polymer electrolyte for high-performance lithium–sulfur batteries. *Journal of Materials Chemistry A*, *7*(22), 13679–13686.
152. Reyes, C., Somogyi, R., Niu, S., Cruz, M. A., Yang, F., Catenacci, M. J., & Wiley, B. J., (2018). Three-dimensional printing of a complete lithium-ion battery with fused filament fabrication. *ACS Applied Energy Materials*, *1*(10), 5268–5279.
153. Chen, Z., Zhao, D., Liu, B., Nian, G., Li, X., Yin, J., & Yang, W., (2019). 3D printing of multifunctional hydrogels. *Advanced Functional Materials*, *29*(20), 1900971.
154. Ndruru, S. T. C. L., Wahyuningrum, D., Bundjali, B., & Arcana, I. M., (2020). Preparation and characterization of biopolymer electrolyte membranes based on liclo4-complexed methyl cellulose as lithium-ion battery separator. *J. Eng. Technol. Sci.*, *52*, 28.
155. Alday, P. P., Barros, S. C., Alves, R., Esperança, J. M., Navarro-Segarra, M., Sabaté, N., & Esquivel, J. P., (2020). Biopolymer electrolyte membranes (BioPEMs) for sustainable primary redox batteries. *Advanced Sustainable Systems*, *4*(2), 1900110.
156. Kim, J. M., Kim, C., Yoo, S., Kim, J. H., Kim, J. H., Lim, J. M., & Lee, S. Y., (2015). Agarose-biofunctionalized, dual-electrospun heteronanofiber mats: Toward metal-ion chelating battery separator membranes. *Journal of Materials Chemistry A*, *3*(20), 10687–10692.
157. Goncalves, R., Lizundia, E., Silva, M. M., Costa, C. M., & Lanceros-Méndez, S., (2019). Mesoporous cellulose nanocrystal membranes as battery separators for environmentally safer lithium-ion batteries. *ACS Applied Energy Materials*, *2*(5), 3749–3761.
158. Liew, S. Y., Walsh, D. A., & Thielemans, W., (2013). High total-electrode and mass-specific capacitance cellulose nanocrystal-polypyrrole nanocomposites for supercapacitors. *RSC Advances*, *3*(24), 9158–9162.
159. Lee, Y. H., Lee, Y. F., Chang, K. H., & Hu, C. C., (2011). Synthesis of N-doped carbon nanosheets from collagen for electrochemical energy storage/conversion systems. *Electrochemistry Communications*, *13*(1), 50–53.

160. Zhou, J., Li, Y., Xie, L., Xu, R., Zhang, R., Gao, M., & Kong, B., (2021). Humidity-sensitive, shape-controllable, and transient zinc-ion batteries based on plasticizing gelatin-silk protein electrolytes. *Materials Today Energy*, *21*, 100712.
161. Zhang, L., Li, H., Lai, X., Gao, T., Liao, X., Chen, W., & Zeng, X., (2019). Carbonized cotton fabric-based multilayer piezoresistive pressure sensors. *Cellulose*, *26*, 5001–5014.
162. Wang, Z., Guo, S., Li, H., Wang, B., Sun, Y., Xu, Z., & Hu, W., (2019). The semiconductor/conductor interface piezoresistive effect in an organic transistor for highly sensitive pressure sensors. *Advanced Materials*, *31*(6), 1805630.
163. Yang, Y., Yang, X., Tan, Y., & Yuan, Q., (2017). Recent progress in flexible and wearable bio-electronics based on nanomaterials. *Nano Research*, *10*, 1560–1583.
164. Liu, M., Pu, X., Jiang, C., Liu, T., Huang, X., Chen, L., & Wang, Z. L., (2017). Large-area all-textile pressure sensors for monitoring human motion and physiological signals. *Advanced Materials*, *29*(41), 1703700.
165. Sun, Q., Qian, B., Uto, K., Chen, J., Liu, X., & Minari, T., (2018). Functional biomaterials towards flexible electronics and sensors. *Biosensors and Bioelectronics*, *119*, 237–251.
166. Hemmati, M., Tafreshi, M. J., Ehsani, M. H., & Alamdari, S., (2022). Highly sensitive and wide-range flexible sensor based on hybrid $BaWO_4$@ CS nanocomposite. *Ceramics International*, *48*(18), 26508–26518.
167. Zhao, S., & Ahn, J. H., (2022). Rational design of high-performance wearable tactile sensors utilizing bioinspired structures/functions, natural biopolymers, and biomimetic strategies. *Materials Science and Engineering: R: Reports*, *148*, 100672.
168. Li, Y., Guo, J., Li, M., Tang, Y., Murugadoss, V., Seok, I., & Luo, Y., (2021). Recent application of cellulose gel in flexible sensing-a review. *ES Food & Agroforestry*, *4*, 9–27.
169. Lu, L., Huang, Z., Li, X., Li, X., Cui, B., Yuan, C., & Dai, Q., (2022). A high-conductive, anti-freezing, antibacterial and anti-swelling starch-based physical hydrogel for multifunctional flexible wearable sensors. *International Journal of Biological Macromolecules*, *213*, 791–803.
170. Bhanu, N., Harikumar, M. E., & Batabyal, S. K., (2022). Self-powered low-range pressure sensor using biopolymer composites. *Applied Physics A*, *128*(3), 214.
171. Zeng, S., Zhang, J., Zu, G., & Huang, J., (2021). Transparent, flexible, and multifunctional starch-based double-network hydrogels as high-performance wearable electronics. *Carbohydrate Polymers*, *267*, 118198.
172. Fu, Q., Cui, C., Meng, L., Hao, S., Dai, R., & Yang, J., (2021). Emerging cellulose-derived materials: A promising platform for the design of flexible wearable sensors toward health and environment monitoring. *Materials Chemistry Frontiers*, *5*(5), 2051–2091.
173. Liu, X., Wei, Y., & Qiu, Y., (2021). Advanced flexible skin-like pressure and strain sensors for human health monitoring. *Micromachines*, *12*(6), 695.
174. Hui, N., Sun, X., Niu, S., & Luo, X., (2017). PEGylated polyaniline nanofibers: Antifouling and conducting biomaterial for electrochemical DNA sensing. *ACS Applied Materials & Interfaces*, *9*(3), 2914–2923.
175. Fan, B., Fan, Q., Cui, M., Wu, T., Wang, J., Ma, H., & Wei, Q., (2019). Photoelectrochemical biosensor for sensitive detection of soluble CD44 based on the facile construction of a poly (ethylene glycol)/hyaluronic acid hybrid antifouling interface. *ACS Applied Materials & Interfaces*, *11*(27), 24764–24770.

176. Huang, W., Ling, S., Li, C., Omenetto, F. G., & Kaplan, D. L., (2018). Silkworm silk-based materials and devices generated using bio-nanotechnology. *Chemical Society Reviews*, *47*(17), 6486–6504.
177. Yu, X., Shou, W., Mahajan, B. K., Huang, X., & Pan, H., (2018). Materials, processes, and facile manufacturing for bioresorbable electronics: A review. *Advanced Materials*, *30*(28), 1707624.
178. Liu, X., Liu, J., Wang, J., Wang, T., Jiang, Y., Hu, J., & Yu, J., (2020). Bioinspired, microstructured silk fibroin adhesives for flexible skin sensors. *ACS Applied Materials & Interfaces*, *12*(5), 5601–5609.
179. Wang, Q., Jian, M., Wang, C., & Zhang, Y., (2017). Carbonized silk nanofiber membrane for transparent and sensitive electronic skin. *Advanced Functional Materials*, *27*(9), 1605657.
180. Zhang, M., Wang, C., Wang, Q., Jian, M., & Zhang, Y., (2016). Sheath–core graphite/silk fiber made by dry-Meyer-rod-coating for wearable strain sensors. *ACS Applied Materials & Interfaces*, *8*(32), 20894–20899.
181. Xu, B., Akhtar, A., Liu, Y., Chen, H., Yeo, W. H., Park, S. I., & Rogers, J. A., (2016). An epidermal stimulation and sensing platform for sensorimotor prosthetic control, management of lower back exertion, and electrical muscle activation. *Advanced Materials*, *28*(22), 4462–4471.
182. Zhong, Q., Zhong, J., Cheng, X., Yao, X., Wang, B., Li, W., & Zhou, J., (2015). Paper-based active tactile sensor array. *Advanced Materials*, *27*(44), 7130–7136.
183. Xu, H., Lv, Y., Qiu, D., Zhou, Y., Zeng, H., & Chu, Y., (2019). An ultra-stretchable, highly sensitive and biocompatible capacitive strain sensor from an ionic nanocomposite for on-skin monitoring. *Nanoscale*, *11*(4), 1570–1578.
184. Hou, C., Xu, Z., Qiu, W., Wu, R., Wang, Y., Xu, Q., & Guo, W., (2019). A biodegradable and stretchable protein-based sensor as artificial electronic skin for human motion detection. *Small*, *15*(11), 1805084.
185. Wang, B., Li, J., Tang, Z., Yang, B., & Lu, S., (2019). Near-infrared emissive carbon dots with 33.96% emission in aqueous solution for cellular sensing and light-emitting diodes. *Science Bulletin*, *64*(17), 1285–1292.
186. Koo, J. H., Kim, D. C., Shim, H. J., Kim, T. H., & Kim, D. H., (2018). Flexible and stretchable smart display: Materials, fabrication, device design, and system integration. *Advanced Functional Materials*, *28*(35), 1801834.
187. Fukuda, K., Takeda, Y., Mizukami, M., Kumaki, D., & Tokito, S., (2014). Fully solution-processed flexible organic thin film transistor arrays with high mobility and exceptional uniformity. *Scientific Reports*, *4*(1), 3947.
188. Liu, Y. F., Feng, J., Bi, Y. G., Yin, D., & Sun, H. B., (2019). Recent developments in flexible organic light-emitting devices. *Advanced Materials Technologies*, *4*(1), 1800371.
189. Gotta, J., Shalom, T. B., Aslanoglou, S., Cifuentes-Rius, A., Voelcker, N. H., Elnathan, R., & Richter, S., (2018). Stable white light-emitting biocomposite films. *Advanced Functional Materials*, *28*(24), 1706967.
190. Jin, J., Lee, D., Im, H. G., Han, Y. C., Jeong, E. G., Rolandi, M., & Bae, B. S., (2016). Chitin nanofiber transparent paper for flexible green electronics. *Advanced Materials*, *28*(26), 5169–5175.

191. Posati, T., Aluigi, A., Donnadio, A., Sotgiu, G., Mosconi, M., Muccini, M., & Seri, M., (2019). Keratin film as natural and eco-friendly support for organic optoelectronic devices. *Advanced Sustainable Systems*, *3*(11), 1900080.
192. Wang, C. Y., Fuentes-Hernandez, C., Chou, W. F., & Kippelen, B., (2017). Top-gate organic field-effect transistors fabricated on paper with high operational stability. *Organic Electronics*, *41*, 340–344.
193. Wang, C. Y., Fuentes-Hernandez, C., Liu, J. C., Dindar, A., Choi, S., Youngblood, J. P., & Kippelen, B., (2015). Stable low-voltage operation top-gate organic field-effect transistors on cellulose nanocrystal substrates. *ACS Applied Materials & Interfaces*, *7*(8), 4804–4808.
194. Pang, C., Koo, J. H., Nguyen, A., Caves, J. M., Kim, M. G., Chortos, A., & Bao, Z., (2015). Highly skin-conformal microhairy sensor for pulse signal amplification. *Advanced Materials*, *27*(4), 634–640.
195. Wang, S., Cheng, H., Yao, B., He, H., Zhang, L., Yue, S., & Ouyang, J., (2021). Self-adhesive, stretchable, biocompatible, and conductive nonvolatile eutectogels as wearable conformal strain and pressure sensors and biopotential electrodes for precise health monitoring. *ACS Applied Materials & Interfaces*, *13*(17), 20735–20745.
196. Chen, X., Jia, B., Zhang, Y., & Gu, M., (2013). Exceeding the limit of plasmonic light trapping in textured screen-printed solar cells using Al nanoparticles and wrinkle-like graphene sheets. *Light: Science & Applications*, *2*(8), e92-e92.
197. Natali, M., Quiroga, S. D., Passoni, L., Criante, L., Benvenuti, E., Bolognini, G., & Toffanin, S., (2017). Simultaneous tenfold brightness enhancement and emitted-light spectral tunability in transparent ambipolar organic light-emitting transistor by integration of high-k photonic crystal. *Advanced Functional Materials*, *27*(21), 1605164.
198. Fujita, M., Ishihara, K., Ueno, T., Asano, T., Noda, S., Ohata, H., & Shimoji, N., (2005). Optical and electrical characteristics of organic light-emitting diodes with two-dimensional photonic crystals in organic/electrode layers. *Japanese Journal of Applied Physics*, *44*(6R), 3669.
199. Chouler, J., Cruz-Izquierdo, Á., Rengaraj, S., Scott, J. L., & Di Lorenzo, M., (2018). A screen-printed paper microbial fuel cell biosensor for detection of toxic compounds in water. *Biosensors and Bioelectronics*, *102*, 49–56.
200. Adekunle, A., Raghavan, V., & Tartakovsky, B., (2019). On-line monitoring of heavy metals-related toxicity with a microbial fuel cell biosensor. *Biosensors and Bioelectronics*, *132*, 382–390.
201. Ferri, J., Llinares, L. R., Moreno, J., Ibañez, C. J., & Garcia-Breijo, E., (2019). A wearable textile 3D gesture recognition sensor based on screen-printing technology. *Sensors*, *19*(23), 5068.
202. Yu, X., Zhou, N., Han, S., Lin, H., Buchholz, D. B., Yu, J., & Facchetti, A., (2013). Flexible spray-coated TIPS-pentacene organic thin-film transistors as ammonia gas sensors. *Journal of Materials Chemistry C*, *1*(40), 6532–6535.
203. Han, S., Jeong, H., Jang, H., Baek, S., Kim, S. H., & Lee, H. S., (2017). Exploring the ultrasonic nozzle spray-coating technique for the fabrication of solution-processed organic electronics. *Organic Electronics*, *49*, 212–217.
204. Mitra, K. Y., Polomoshnov, M., Martínez-Domingo, C., Mitra, D., Ramon, E., & Baumann, R. R., (2017). Fully inkjet-printed thin-film transistor array manufactured on paper substrate for cheap electronic applications. *Advanced Electronic Materials*, *3*(12), 1700275.

205. Chung, S., Cho, K., & Lee, T., (2019). Recent progress in inkjet-printed thin-film transistors. *Advanced Science*, *6*(6), 1801445.
206. Yadav, B. S., Dey, S. R., & Dhage, S. R., (2019). Effective ink-jet printing of aqueous ink for Cu (In, Ga) Se_2 thin film absorber for solar cell application. *Solar Energy*, *179*, 363–370.
207. Zimmermann, J., Schlisske, S., Held, M., Tisserant, J. N., Porcarelli, L., Sanchez-Sanchez, A., & Hernandez-Sosa, G., (2019). Ultrathin fully printed light-emitting electrochemical cells with arbitrary designs on biocompatible substrates. *Advanced Materials Technologies*, *4*(3), 1800641.
208. Conti, S., Martínez-Domingo, C., Lay, M., Terés, L., Vilaseca, F., & Ramon, E., (2020). Nanopaper-based organic inkjet-printed diodes. *Advanced Materials Technologies*, *5*(6), 1900773.
209. Kang, H. W., Lee, S. J., Ko, I. K., Kengla, C., Yoo, J. J., & Atala, A., (2016). A 3D bioprinting system to produce human-scale tissue constructs with structural integrity. *Nature Biotechnology*, *34*(3), 312–319.
210. Zarek, M., Layani, M., Cooperstein, I., Sachyani, E., Cohn, D., & Magdassi, S., (2016). 3D printing of shape memory polymers for flexible electronic devices. *Advanced Materials*, *28*(22), 4449–4454.
211. Muth, J. T., Vogt, D. M., Truby, R. L., Mengüç, Y., Kolesky, D. B., Wood, R. J., & Lewis, J. A., (2014). Embedded 3D printing of strain sensors within highly stretchable elastomers. *Advanced Materials*, *26*(36), 6307–6312.
212. Zhang, M., Zhao, M., Jian, M., Wang, C., Yu, A., Yin, Z., & Zhang, Y., (2019). Printable smart pattern for multifunctional energy-management E-textile. *Matter*, *1*(1), 168–179.
213. Wang, Ziying & Ma, Zongtao & Sun, Jingyao & Yan, Yuhua & Bu, Miaomiao & Huo, Yanming & Li, Yun-Fei & Hu, Ning. (2021). Recent Advances in Natural Functional Biopolymers and Their Applications of Electronic Skins and Flexible Strain Sensors. Polymers. 13. 813. 10.3390/polym13050813.
214. Peters, J.F.; Peña Cruz, A.; Weil, M. Exploring the Economic Potential of Sodium-Ion Batteries. Batteries 2019, 5, 10. https://doi.org/10.3390/batteries5010010
215. Dalwadi, Shrey & Goel, Arnav & Kapetanakis, Constantine & Cruz, David & Hu, Xiao. (2023). The Integration of Biopolymer-Based Materials for Energy Storage Applications: A Review. International Journal of Molecular Sciences. 24. 3975. 10.3390/ijms24043975.

CHAPTER 3

Plant Fiber-Based Bio-Hybrid Composites: Green Technologies for a Sustainable Environment

ROLI VERMA,[1] DHIRAJ DUTTA,[2] and RAMA DUBEY[2]

[1]*Department of Chemistry, JSS Academy of Technical Education, Noida, Uttar Pradesh, India*

[2]*SPD Division, Defense Research Laboratory, DRDO Tezpur, Assam, India*

ABSTRACT

Hybrid materials are quickly becoming a powerful and exciting class of materials due to the need to adapt to later advancements like industrial ecology, green chemistry, and sustainable development. New materials are produced. Among this new group of materials are composites with exceptional performance levels. Numerous studies are looking into how bio-composites have developed to produce more useful and economical things because petroleum and other conventional resources seem to be depleting and are pricey. Additionally, it refers to "natural/synthetic fibers" or "natural fibers" in bio-hybrid composite materials. The overall synergy of the properties of the needed material results in the production of the hybrid final product. The variety of the resulting features and materials used to create hybrid materials benefits applications in the automotive, structural, electrical, and biological fields. Currently, the most advanced and high-tech engineering materials are those that are multifunctional. They have superior chemical and environmental resistance, high thermal stability, fatigue resistance, and the capacity to endure mechanical loads. These materials with natural reinforcement have a high potential to replace conventional composites due to their low cost, light

Sustainable Environmental Engineering: Water Security, Energy Conservation, and Green Process.
Sonia Khanna, Ann Rose Abraham, Kian Hariri Asli, & A. K. Haghi (Eds.)

weight, and favorable environmental effects. Furthermore, it offers details on the origins of natural reinforcements, their mechanical and chemical properties, and the environments in which they may be found. In this chapter, we highlight the research that has been done on hybrid materials, discuss their wide range of uses, and provide a prediction for their future.

3.1 INTRODUCTION TO HYBRID COMPOSITE MATERIALS

To improve the qualities of each individual part, composite materials are created by combining two or more easily distinct elements [1, 2]. There are many reasons why newly developed materials may be preferred, including the fact that they are more affordable, lighter, and stronger than currently used materials [3–5]. Constituents are the collective term for the various components that make up composites. However, in general, the composites may contain not only two components, matrix and reinforcement, but also other types of components: fillers, compatibilizers, coupling agents, pigments, lubricants, surfactants, solvents, etc. Most composites primarily consist of two constituent materials: reinforcement and matrix. The reinforcement, which can be natural, synthetic, or hybrid fibers or fabrics, and the polymer matrix are the only two components found in even the most basic textile-based composites, also known as textolites. The improved properties of composites are a result of the reinforcement, which is substantially stiffer and stronger than the matrix [6, 7]. In textile-based composites, the polymer matrix's primary jobs are to bind reinforcements (such as fabric, fibers, or nanofibers) and preserve the composite's structural integrity. The field of nanoscience, however, only opened up many possibilities for approaches to new materials at the end of the 20th and the beginning of the 21st centuries as a result of the availability of novel physico-chemical characterization techniques. Combining various analytical methods resulted in new insights into hybrid materials and demonstrated that bottom-up approaches to material design that start at the molecular level will produce novel properties in this class of materials.

Figure 3.1 illustrates the advancements in composite structures from conventional composites to Green composites.

Environmentally friendly materials have come into consideration as a result of rising environmental awareness and public interest, new environmental regulations, and unsustainable petroleum usage. When compared to synthetic fiber, natural fiber [8–12] is thought of as one of

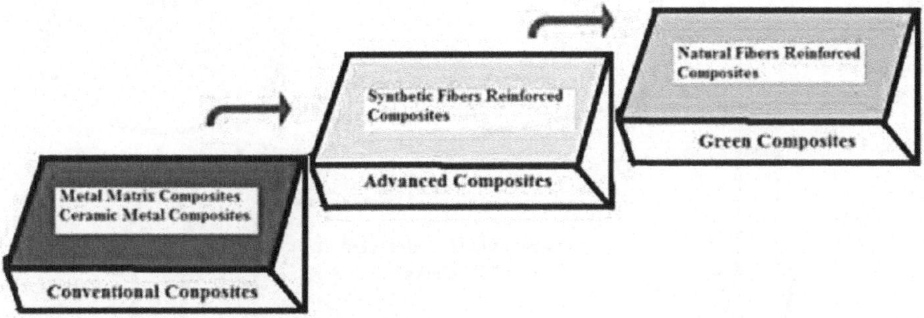

FIGURE 3.1 Advancement in composite structures.

the ecologically friendly materials with superior qualities [14]. Concerns about the environment and sustainability [15] as well as a desire to catch up with cutting-edge concepts like green chemistry and the green industrial revolution have sparked the creation of bio-composites. Contrary to composites made with synthetic reinforcement, biomaterials can be discarded and converted into ecological compost at the end of their useful lifetimes. These green materials have generated a lot of interest and exhibit great promise for a range of uses, including the construction, automotive, and sports components industries, among others, as a result of their improved qualities. It is also notable that the proportion of these renewable raw materials used in the production of biomaterials has increased lately, from 5% in 2004 to 13% in 2010 and 18% in 2020. Predictions state that by 2030, it will be 25%. Numerous bio-hybrid composites have recently been developed using a range of organic and inorganic reinforcing materials. Because of characteristics like renewability, low density, accessibility, and affordability, natural fibers are an alternative to synthetic fibers. In addition to having properties that alternate between inorganic and organic materials, these materials also exhibit novel behaviors that allow them to be used in a wider variety of applications [13]. By adding natural fibers to a variety of matrices, including thermoplastics (Polypropylene, Polyethylene, etc.) and thermosets, numerous studies have been conducted to determine the mechanical and final characteristics of biocomposites/bio-hybrid composites (polyester, epoxy, etc.). All available information indicates that a critical element influencing the efficacy of the reinforcement is the interfacial adhesion of the matrix and fibers (Figure 3.2).

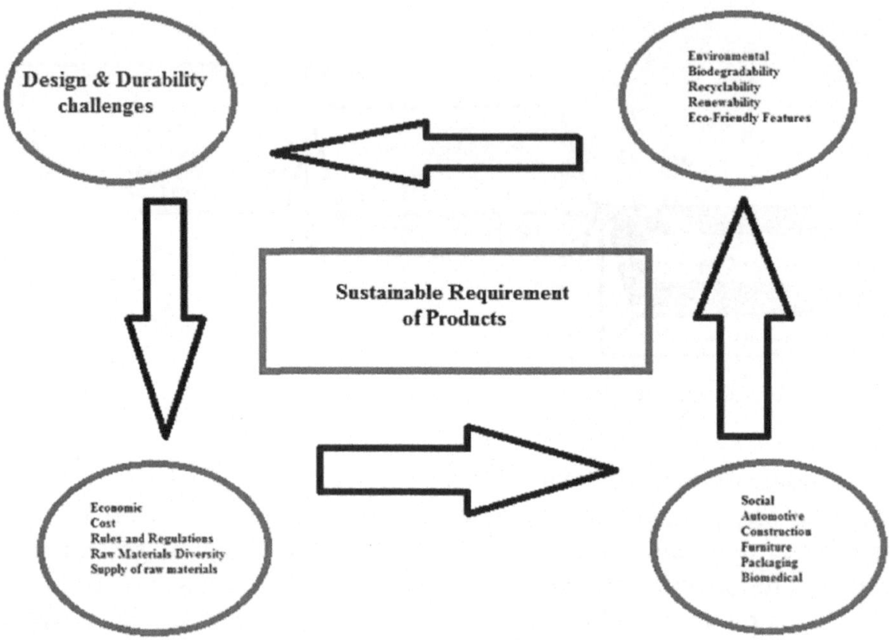

FIGURE 3.2 Parameters associated with sustainable products.

3.1.1 CLASSIFICATION OF HYBRID COMPOSITES

According to the potential interactions between the organic and inorganic species contained in the material, hybrid materials are categorized. Van der Waals, hydrogen bonds, and weak electrostatic interactions between the two constituents are weak in Class I materials. Class II materials have robust chemical interactions between the two constituents, or covalent bonding. Generally, five main types of hybrid composite materials are assumed to be distinguished (Figure 3.3):

1. **Averaged:** Fibers of different types in such a PHCM are mixed in the whole mass of the material. No sections with macro-concentrations of any type of fiber are available.
2. **Intralaminar:** Fibers of different types in each layer of PHCM are regularly alternated. Referred to this type of PHCM are materials based on reinforcing hybrid fabrics. The layers can be 'interwoven.'
3. **Interlaminar:** Each layer of PHCM consists of one type of fiber. As a whole, the PHCM is made up of layers alternating in a definite

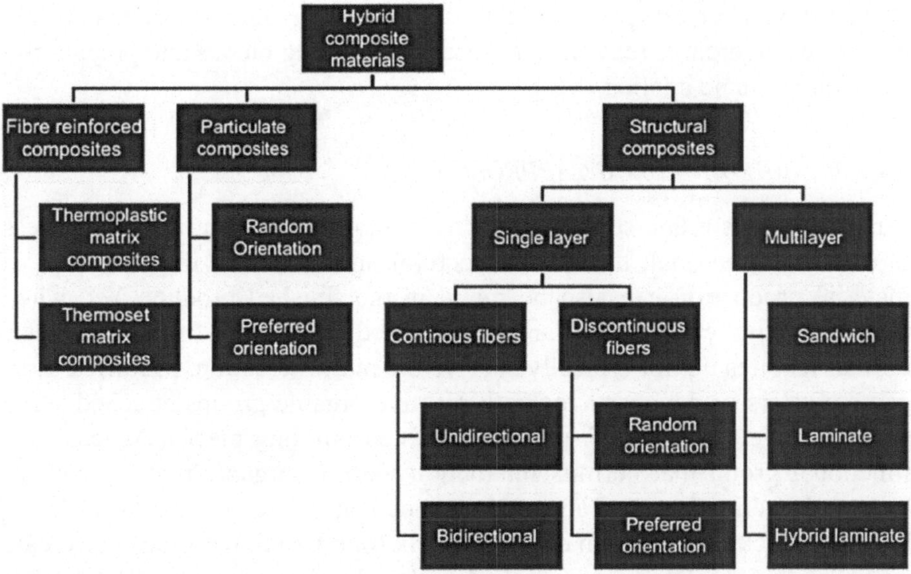

FIGURE 3.3 Classification of hybrid composites.

Source: Reprinted with permission from Ref. [63].

order from fibers of different types (e.g., a part from carbon plastic or organic fiber-reinforced plastic).

4. **Separate Reinforcing Elements:** These include straps and stiffening ribs.
5. **Super Hybrids:** A specific pattern is followed while layering composite materials made of organic polymers, metal matrix, and sheets of metal foil. Most frequently employed in the building of medium-loaded and heavily-loaded aircraft are intralaminar and interlaminar PHCMs based on carbon, glass, and organic fillers.

3.1.2 SYNTHESIS OF HYBRID COMPOSITE MATERIALS

To create hybrid materials, one of two methods can be applied: In order to create the final hybrid material, in which the precursors still retain at least some of their original integrity, either well-defined premade building blocks are used, or one or both structural units are generated from the precursors and are changed into a new (network) structure. The interface between organic and inorganic materials must be specifically designed to avoid major obstacles when creating hybrid materials. To bridge the disparities between inorganic and organic materials, a variety of building blocks and preparation techniques can be applied.

3.1.2.1 BUILDING BLOCK APPROACH

Building blocks retain some of their molecular integrity during the process of creating a material; therefore, structural units found in these sources of material production can also be found in the finished product. Yet, when these building components are transformed into novel materials, their normal features do not typically survive the matrix creation. Modified inorganic clusters or nanoparticles with reactive organic groups attached serve as illustrative instances of such well-defined building pieces. At least one functional group that permits interaction with an organic matrix, such as by copolymerization, is frequently included in cluster compounds. These building blocks can alter an organic matrix (one functional group) or create partially or completely cross-linked materials, depending on the number of groups that can interact (more than one group). Chain structures, for instance, can develop when two reactive groups interact. The creation of a cross-linked material can be accomplished using only the building blocks if they have at least three reactive groups.

Besides the molecular building blocks described, nano-sized building blocks, such as particles or nano-rods, can also be utilized to make nanocomposites. The building block approach has one significant advantage over the *in situ* formation of the inorganic or organic entities: better structure-property predictions are possible because at least one structural unit (the building block) is well-defined and typically does not undergo significant structural changes during the matrix formation.

Moreover, building blocks can be developed to offer the best performance when producing materials. The solubility of inorganic compounds in organic monomers, for instance, can be increased by surface groups having similar polarity as the monomers. Several building blocks have been synthesized recently and used to create hybrid materials. These substances can be created on a molecular level by chemists using extremely complex techniques, and the resulting systems are used to create functional hybrid materials. Future applications, particularly in nanotechnology, frequently take a bottom-up strategy in which simple building components are arranged in hierarchical fashion to create complex structures. The building block technique in hybrid materials is also motivated by this concept.

3.1.2.2 IN SITU FORMATION OF THE COMPONENTS

The chemical transformation of the precursors utilized during the manufacturing of the materials forms the basis for the *in situ* formation of the hybrid materials. Usually, this happens when organic polymers are created, but it can also happen when an inorganic component is created via the sol-gel method. In these situations, discrete, well-defined molecules are converted into multidimensional formations that frequently exhibit entirely different properties than their precursors. Typically, straightforward, easily accessible molecules are used, and the internal structure of the final product is influenced both by the makeup of these precursors and the reaction conditions. Control of the latter is therefore an essential element in this procedure. Frequently, altering one parameter results in two quite distinct materials. The difference between base and acid catalysis is significant if the inorganic species is, for instance, a silica derivative produced by the sol-gel method because base catalysis results in a more particle-like microstructure while acid catalysis results in a more polymer-like microstructure. As a result, the processing of the obtained materials and its optimization have a significant impact on their final performance.

3.1.2.3 IN SITU FORMATION OF INORGANIC MATERIALS

As organic groups are pronc to decomposition at high temperatures, many of the traditional inorganic solid-state materials are created using solid precursors and high-temperature methods, which are frequently incompatible with the presence of organic groups. Thus, the *in situ* synthesis of hybrid materials is not appropriate for these high-temperature procedures. The reactions used ought to be more akin to the way that covalent bonds form in solutions traditionally. The sol-gel process is one of the most well-known methods that satisfies these requirements. However, these very low-temperature reactions frequently result in kinetic products rather than the thermodynamically most stable structure, which has some ramifications for the structures formed. For instance, inorganic materials produced at low temperatures are frequently amorphous or only exhibit crystallinity at very small length scales, such as the nanometer range. The latter is exemplified by the reduction of metal salts or organ metallic precursors to produce metal nanoparticles in organic or inorganic matrices. Some methods of *in situ* formation of inorganic materials are given below.

- Sol-gel process;
- Nonhydrolytic sol–gel process;
- Sol–gel reactions of non-silicates.

3.1.2.4 FORMATION OF ORGANIC POLYMERS IN PRESENCE OF PREFORMED INORGANIC MATERIALS

To overcome the incompatibility of the two species, one must choose between a number of options if the organic polymerization takes place in the presence of an inorganic substance to create the hybrid material. The surface of the inorganic material may be left unmodified, leaving only the naked surface, modified with nonreactive organic groups (such as alkyl chains), or it may contain reactive surface groups like polymerizable functions. In accordance with these requirements, the material may need to be pretreated. For instance, a pure inorganic surface may be treated with surfactants or silane coupling agents to make it compatible with the organic monomers, or functional monomers may be added that interact with the inorganic material's surface. The final product of the organic polymerization is a blend if the inorganic component contains nonreactive organic groups attached to its surface and can be dissolved in a monomer that is then polymerized. In this instance, the inorganic component interacts with the organic polymer only slightly or not at all, leading to the formation of a class I material. Only if aggregation of the inorganic components in the organic environment is avoided will homogeneous materials be produced in this scenario. If the interactions between the inorganic components and the monomers are better or at least equivalent to those between the inorganic components, this can be accomplished. The long-term stability of a once homogeneous material, however, is debatable if strong chemical connections do not emerge due to diffusion effects in the resulting hybrid material.

The stronger the respective interaction between the components, the more stable the final material is. The strongest interaction is achieved if class II materials are formed, for example with covalent interactions.

3.1.2.5 HYBRID MATERIALS BY SIMULTANEOUS FORMATION OF BOTH COMPONENTS

The most homogeneous type of interpenetrating networks can be produced by simultaneously producing inorganic and organic polymers. Typically, the organic polymerization monomers and sol-gel process precursors are combined, and both processes— with or without solvent— are carried out simultaneously. Three processes are in competition with one another when using this method:

- the kinetics of the hydrolysis and condensation forming the inorganic phase;

- the kinetics of the polymerization of the organic phase; and
- the thermodynamics of the phase separation between the two phases.

Phase separation is prevented or reduced by adjusting the kinetics of the two polymerizations so that they occur concurrently and quickly enough. Phase separation can also be avoided by using additional parameters, such as the favorable interactions between the two moieties described above.

The sensitivity of many organic polymerization methods to sol-gel conditions or the makeup of the materials generated is another issue brought on by the concurrent creation of both networks. For instance, ionic polymerizations frequently interact with the sol-gel process's precursors or intermediates. As a result, they are rarely used in these reactions.

3.2 NATURAL FIBER REINFORCED POLYMER COMPOSITES (NFRPCS)

Natural fiber reinforced polymer composites (NFRPCs) are considered one of the emerging materials of contemporary times, and the term "green" is often used to describe these NFRPCs [17]. This is due to the disintegration of the various components of composite materials or, in other words, the biodegradability of NFRPCs. Sustainability, recyclability, and biodegradability can have a significant impact on both the current and future climate. Environmentally friendly products are gaining popularity due to increasing regulations and legislation against harmful products. In this context, researchers are promoting the production of green materials, particularly NFRPCs [18–21]. NFRPCs are largely produced thanks to NFs. Additionally, the manufacturing of NFRPCs uses less energy (9.55 MJ/kg) than traditional fiber-reinforced composites, such as glass (54.7 MJ/kg). Environmentally speaking, NFRPCs are less harmful than synthetic fiber reinforced composites. These products with economical features like biodegradability and renewability are gaining popularity because they have less of an impact on the environment [22].

The money made from the production of natural fibers is another important advantage connected with NFs. Additionally, it is possible to repeatedly cultivate the ground used for NF creation. For example, the production of hemp and flax fibers results in the production of seeds, chemicals, and oils that have a variety of essential applications, including those as dietary supplements for people. At the end of its life cycle, the material produced can also decompose. An industry that generates 64.3 billion nuts annually, for example, already creates coir strands as a byproduct. The majority of the

NFs in these NFRPCs range from 60 to 70%, with the glue and matrix making up the remainder. In an open setting, environmental factors that influence NF degradation include humidity, temperature, ultraviolet (UV) light, and the presence of various microorganisms. The cellulose, lignin, and hemicellulose of the fibers break down during degradation, completely ruining the NFRPCs' mechanical qualities. The tensile strength of the kenaf/POM composites was lower after being exposed to moisture, a water shower, and UV radiation in a research [23]. This result was ascribed to the kenaf strands' cellulose, hemicelluloses, and lignin degrading. It was investigated how the atmosphere affected the biodegradability of jute/phenolic compounds. According to the findings, after two years of UV exposure, almost half of the elastic properties were reduced. Due to the UV exposure, pitch breaking, swelling, fibrillation, and black patches were also noted.

3.3 DIFFERENT TYPES OF NATURAL FIBERS

Simple natural fabrics that are not synthetic or man-made. Both plants and mammals contain natural fiber [24]. Natural fiber that comes from both types of materials can be used. Natural cellulose fiber comes in a wide variety of forms, including flax, hemp, sisal, banana, kenaf, jute, and fruit cluster cellulose from oil palm trees. Jute fiber-based composite materials have attracted a lot more notice in recent years. The plants that produce cellulose fibers can be divided into bast fibers, such as jute, flax, ramie, hemp, and kenaf, seed fibers, such as cotton, coir, and kapok, leaf fibers, such as sisal, pineapple, and abaca, grass and reed fibers, such as rice, corn, and wheat, and core fibers, such as hemp, kenaf, and jute (wood and roots). The empty fruit bunch cellulose fibers are difficult to dispose of in Southeast Asia, especially Indonesia and Malaysia, which are top producers of palm oil, so it is important to turn the waste into usable reinforcement. Due to their low density, renewability, biodegradability, non-toxicity, excellent insulation properties, and resistance to machine wear, natural fibers are advantageous as plastic reinforcement. The low density of natural fibers is particularly beneficial for the automotive sector [25].

The two main types of fibers are natural and synthetic fibers. Natural fibers are those produced organically [26]. Synthetic fibers are those created artificially. Plant fibers are further categorized into primary fibers, such as jute, sisal, hemp, and others grown mainly for their fiber content, and secondary fibers, which are leftovers from plants. Natural fibers encompass raw materials resembling hair derived directly from plant, animal, mineral, and vegetable sources. These raw materials are processed into nonwoven fabrics, filaments, threads, ropes, and can be used in composite materials and paper production. Many researchers have shown interest in natural fibers due to their abundance in nature. They explore ways to expand the use of natural fibers and improve the properties of composite materials containing them. The categorization of natural fibers is illustrated in Figure 3.4.

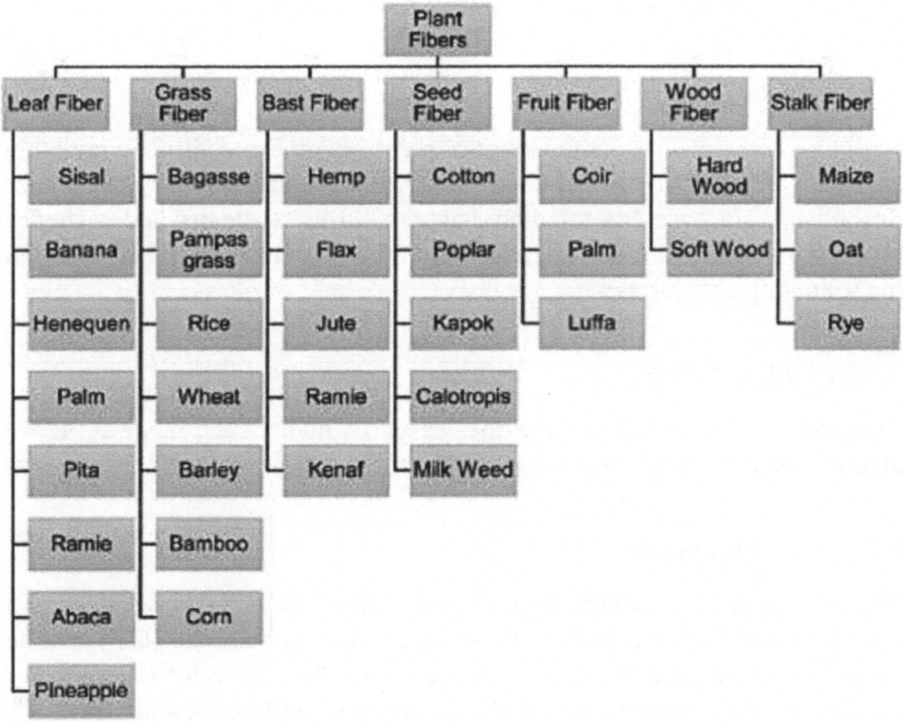

FIGURE 3.4 Classification of natural fibers.

There are natural fabrics made from plant, animal, and mineral sources. Natural fibers can be categorized into three groups according to their sources: animal fibers, mineral fibers, and plant cellulose fibers.

3.3.1 ANIMAL FIBERS

Animal fibers include fibers from alpacas, birds, wool, silk, hair, and fur. Sheep's wool, goat hair (including alpaca fiber and mohair from Angora goats), rabbit's wool, Siberian weasel, ox hair, hog bristle, camel hair, horse-hair, feathers, and feather fiber are all included.

3.3.2 MINERAL FIBER

Mineral fibers are derived from nature or, in some cases, require minor modification to become asbestos, ceramic, or metal fibers. Serpentine, amphiboles, and anthophyllite are examples of naturally occurring mineral fibers that are similar to asbestos. Glass fibers, aluminum oxide, silicon carbide, and boron carbide are examples of ceramic fibers, and aluminum fibers are an example of mineral fibers.

3.3.3 PLANT FIBER

Cellulose is the primary component of plant fibers. Depending on how they are used, plant fibers are divided into primary and secondary categories. Secondary plants are those that are harvested from waste products, whereas primary plants are farmed for their fibers. This fiber can also be divided into the following categories:

3.3.3.1 LEAF FIBER

These fibers are collected from the leaves of plants such as sisal, agave, abaca, henequen, and pineapple.

3.3.3.1.1 Pineapple

The pineapple is a natural tropical plant that was initially discovered in Brazil, South America, and the West Indies. Today, there are many distinct varieties of pineapple plants that are used as a fabric for industrial purposes, medicines, edibles, textile materials, the production of yarns, and handicrafts in many different nations [27, 28]. Compared to jute fiber, PALF has a higher mechanical strength, superior chemical compositions, high specific strength, rigidity, and flexural and high torsional rigidity. Industries can employ PALF as a raw material to create reinforcing composite matrices.

3.3.3.2 SKIN FIBER OR BAST FIBER

These fibers are extracted from the skin or bast that covers the plant stem. The tensile strength of skin or bast fibers is higher than that of other fibers. Because of this, these fibers are employed in the production of paper, cloth, and durable yarn, such as vine fibers, soybean fiber, banana fiber, hemp, kenaf, ramie, rattan, and flax.

3.3.3.2.1 Jute Fiber

Jute (*Corchorus capsularis*/*Corchorus olitorius*) is one of the cheapest natural fibers and the second-most important fiber after cotton fiber in terms of production. It is primarily farmed for its fiber. Jute of the *Corchorus capsularis* and *Corchorus olitorius* types are both found in Africa and Indo-Burma, including South China. Jute's history dates back to 206 BC and ends in 221 AD. Jute paper from the Western Han Dynasty was discovered in Dunhuang, Gansu Province, China. According to historical records, jute fiber was utilized in India beginning in the 16th century, under the rule of Mughal Emperor Akbar. From India, the British East India Company exported jute. Following then, the jute fiber market expanded quickly all over the world, especially in Dundee, Scotland in 1800. Jute is grown in a number of nations, including Bangladesh, India, Myanmar, Nepal, China, Taiwan, Thailand, Vietnam, Cambodia, and Brazil. Bangladesh, India, and China are among them where jute may grow best. Jute fiber has excellent thermal and mechanical performance, a high strength-to-weight ratio, a high aspect ratio, and outstanding insulating qualities [29]. A door, window, piece of furniture, an automobile, an aircraft, water pipes, fake roofing, floor tiles, etc., have all been made from jute fiber-based composite. Subash et al. talked about green composites with bast fiber reinforcement that are applied to aircraft interior structures. The aerospace and automotive industries employ good materials made from natural fibers like jute, kenaf, bamboo, bagasse, coir, and sisal [27, 31]. Natural fiber can be used to create sustainable goods that can significantly lower the amount of energy used in the aerospace sector due to its low density and lightweight characteristics. Jute fiber is utilized in a variety of products, including bags for packaging, carpet backing, ropes, doors, furniture, floor tiles, yarns, and wall art. Jute fiber is a natural fabric with various benefits, including high tensile strength, superior water resistance, biodegradability, and environmental friendliness [32–34].

3.3.3.2.2 Kenaf Fiber

The bast (outer) and core (inner) fibers of the kenaf plant, which shares certain traits with jute, are used to make kenaf fiber (*Hibiscus cannabinus*). Mostly found in Africa and Asia, kenaf fiber. Within 150 days, a kenaf plant can reach a height of 2.4–6 m. The United States grows a variety of kenaf plants whose cultivation is entirely dependent on the weather [67]. They use kenaf fiber for commercial purposes. It has many benefits, including use in the manufacture of composites, the production of twine, rope, sackcloth, paper, animal foods, the production of oil, medicines, and food additives, as well as industrial applications as a platform for mushroom farming, environmental cleaning, absorbents, the automotive industry, textiles, and chemical and oil absorbents. Webber III clarified that the kenaf plant has a variety of helpful components, such as stalks, leaves, and seeds, and that these three components can be further broken down into a variety of useful parts, including fibers and fiber strands, oils, proteins, and allelopathic substances. Many variables, including photosensitivity, plant populations, growing season, planting date, cultivar, and plant maturity, might have an impact on kenaf fiber. Its fiber characteristics are influenced by the sources, age, and separating method. Low density, strong mechanical characteristics, and biodegradability are all features of kenaf fiber [33].

3.3.3.2.3 Flax

Flax is a cellulosic natural fiber that is also a bast fiber. Flax fiber is grown in temperate regions and taken from the epidermis of its own plant's stem before being made into linen yarn. It is one of the oldest fibers and is now primarily employed in a variety of composite applications. Flax fiber is supple and glossy, has exceptional abrasion resistance, and absorbs and dries fast. Because flax fiber is two to three times stronger than cotton fiber, the value of flax fiber reinforced composites is increasing in textile markets. Further benefits of flax fibers include low density, high strength and stiffness, high toughness, and biodegradability. In the production of rope, twine, canvas, and webbing equipment as well as tablecloths, garments, surgical thread, decorative fabrics, and more, flax fiber is utilized as a raw material.

3.3.3.2.4 Hemp

Another significant component of the oldest and most environmentally beneficial fiber family, hemp is a plant that thrives in temperate areas. Excellent mechanical strength and Young's modulus are always provided by hemp fibers. Textile, paper, rope, and oil manufacturing industries all employ hemp fiber [35]. Asia was where it first saw use. Today, hemp fiber is used in non-food agriculture in the European Union. Hemp fiber outperforms cotton fiber in terms of strength, absorbency, resistance, high tensile strength, high luster, and durability. Due to its resistance to water rot, hemp fiber is used to make ship cordage as well as rugs and carpets. Hemp fiber is used to screen the sun's damaging UV radiation as well as for clothing materials and interior design. According to recent research, hemp fiber accounts for one-third of China's entire production. The use of hemp fiber in the textile industry is still somewhat restricted.

3.3.3.3 FRUIT FIBER

These fibers are collected from the fruit of the plant, such as coconut (coir) fiber.

3.3.3.3.1 Coir Fiber

Coir comes from the husk of the coconut fiber, which has a long life compared to other fibers due to its lignin properties. Coir fiber-reinforced composites [36] can be fabricated with both thermoset and thermoplastic resins [37]. The mechanical properties of the coir composite totally depend on its interfacial adhesion characteristics. Coir fiber reinforced composites are manufactured for social, economic, and industrial applications such as packaging material, helmets, rope, finishing nets, brushes, mattresses, automotive interior, paneling, roofing as building materials, postboxes, mirror casing, storage tank, projector cover, and voltage stabilizer cover. However, it has some limitations such as high moisture content, which can be treated using chemical treatment [38–40]. The mechanical properties of the untreated coir-based composite were inferior to those of the pretreated coir-based composite [41]. Jute-coir hybrid composites are employed in the construction of doors, windows, and the interiors of buildings in addition to the transportation industry.

3.3.3.4 STALK FIBER

Fibers are found from the stalks of the plants such as straws of wheat, rice, barley, maize, rye, and oat.

3.3.3.4.1 Sisal

The Agave sisalana plant yields sisal, which is manufactured in Brazil and East Africa. The leaves of the plant, which resemble enormous pine-apples, are stripped away during harvest, leaving the fibrous plant material behind. Soft tissue is then removed from the fibers [34] either manually or mechanically. The residual dirt is then scraped off the fibers after they have dried and gone through this process. Sisal produces strong fibers that are completely biodegradable, robust, abrasion resistant, high tenacity, high tensile strength, acid and alkali resistant, corrosion resistant, and sea water resistant. However, sisal fiber is unsuitable for finishing walls because it is too rough for use in textiles and fabrics. Instead, it is mostly utilized for mats, roofs, carpets, disc buffers, cement reinforcement, and the automotive industry. However, since 1998–2000, the demand for sisal fiber products has been declining globally. Sisal fibers have excellent sound and shock absorption qualities, are long-lasting, recyclable, and require little upkeep [43].

3.3.3.4.2 Bamboo

Natural bamboo is used to create cellulose fibers known as bamboo fibers [44–45]. A mature bamboo plant may reach a height of 40 meters, and it takes the plant six to eight months to attain that height [46]. The global market has recently begun to pay more attention to bamboo fibers. There are over 1,000 known species of bamboo globally. Bamboo grows quite quickly. Its fiber is superior to cotton fibers in many ways. Raw bamboo fiber is utilized in the textile industry. It is also employed in the production of "Meso" fabric. Bamboo fiber is completely eco-friendly and has special qualities like UV resistance, moisture absorption, flexibility, and antibacterial properties. Bamboo fiber has inherent antibacterial properties, which is why it is employed in construction. Bamboo fiber is used to make hygiene, decorative, and bathroom products.

3.3.3.4.3 Abaca

The plant's sturdy, seawater-resistant stalk is where the abaca gathers. Abaca fiber [46] has a number of benefits, including low density, the right stiffness and mechanical qualities, high disposability and renewability, and the ability to be recycled and biodegraded. One of the strongest cellulose fibers, abaca fiber is resistant to saltwater and is frequently used in fishing nets, meat casings, and the manufacture of tea bags. It is also used as a raw material in the production of high-quality paper, machinery filters, hospital textiles, electric conductors, and other products.

3.3.3.5 CANE, GRASS, AND REED FIBER

Fibers are collected from its original nature source, e.g., bamboo, bagasse, sabai, communis, and phragmites (Figure 3.5).

(Continued)

 Jute Plant	 Jute Fiber
 Bamboo Plant	 Bamboo Fiber
 Sisal Plant	 Sisal Fiber
 Coir Plant	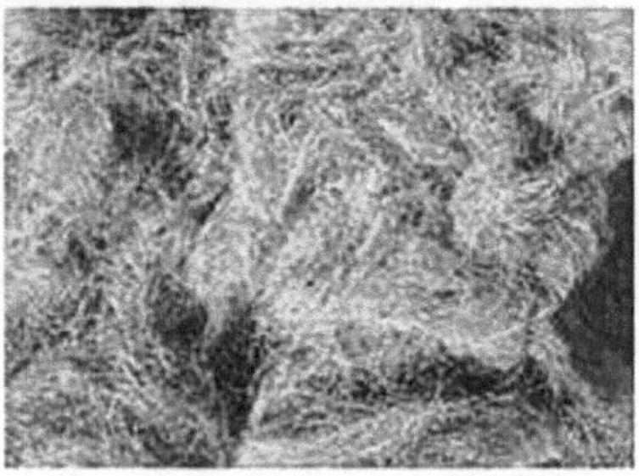 Coir Fiber

(Continued)

FIGURE 3.5 Plant source and their fibers.

3.4 ADVANTAGES OF USAGE OF NATURAL FIBER

Natural fiber has an increasingly greater advantage over synthetic fiber due to its abundance, low cost, lighter weight, low manufacturing cost, low environmental impact, and relatively good physical and mechanical characteristics, including tensile strength, tensile modulus, bending strength, and biodegradability. Particularly in rural developing nations, natural fibers have been grown and used to make non-structural products including bags, brooms, fishnets, and filters. Natural fibers can also be utilized as insulation for walls and roofs. The characteristics of natural fibers differ greatly from one another. The main advantages offered by using natural fiber are as follows:

- Reduction in density of the products is less compared to synthetic fibers. Density of fibers is less and it reduces the weight of the composites.
- Natural fibers are cheaper than synthetic fibers. Natural fibers are non-toxic in nature.
- The manufacturing process of natural fiber reinforced composites is safer than that of synthetic fiber composites.

- The waste produced during the manufacturing of natural fiber composites can be recycled. In automobiles, the fuel consumption is due to less weight of composites.

3.5 PHYSICAL AND MECHANICAL PROPERTIES OF NATURAL FIBER COMPOSITES

To achieve the desired mechanical properties of natural fiber reinforced composites, several factors need to be taken into account. These factors include the type of natural fibers, surface chemistry between fibers, moisture content of fibers, form of fibers, composition (e.g., epoxy resin), and the quality of the interface. The qualities of natural fibers are affected by various factors such as the varieties of natural fibers, their cultivation, growth season, retting or extracting procedure, isolation, and processing method, and other factors. Numerous studies have been conducted to determine the effects of fiber types and processing techniques on the density, elongation at break, Young's modulus, and tensile strength of natural fiber composites. Natural fibers, particularly flax, abaca, kenaf, ramie, hemp, jute, sisal, bamboo, and bagasse, exhibit excellent mechanical qualities and have higher strength and modulus compared to synthetic fibers. Table 3.1 shows the types and mechanical characteristics of the most prevalent natural fibers. Additionally, natural fibers have a low density and a good Young's modulus.

TABLE 3.1 Chemical Composition of Natural Fiber

Fiber	Cellulose (wt.%)	Hemicellulose (wt.%)	Lignin (wt.%)	Tensile Strength (MPa)
Flax	71	18.6–20.6	2.2	–
Kneaf	72	20.3	9	283–800
Jute	61–71	14–20	12–13	393–800
Bamboo	26–43	30	21–31	–
Bagase	55.2	16.8	25.3	
Hemp	68	15	10	550–900
Abaca	56–63	20–25	7–9	–
Sisal	65	12	9.9	–
Coir	32–43	0.15–0.25	40–45	–
Pineapple	81	–	12.7	126.0

The mechanical properties [47] of the composites are also impacted by the stacking order. Improved mechanical qualities are produced by better fiber and matrix stacking. Mechanical qualities should be improved with fiber content increases up to a particular point below the weight percentage of the matrix [48]. Improved fiber and matrix stacking results in materials with better mechanical properties. Increased fiber content in the composite enhances the mechanical characteristics. The mechanical characteristics of composites are also influenced by the fiber's shape. Because natural fibers naturally absorb moisture, composites reinforced with them cannot be used in outdoor applications. The mechanical qualities of natural fibers are inferior to those of synthetic fibers. One method for improving the mechanical properties of composites reinforced with natural fibers is hybridization [49].

Fiber content should be increased up to a specific degree below the matrix's weight percentage in order to improve mechanical properties. The mechanical qualities of the composite are enhanced by adding more fiber. The form of the fiber has an impact on the mechanical properties of composites as well. Composites bonded with natural fibers cannot be used outside because they naturally absorb moisture. Natural fibers' mechanical properties fall short of those of synthetic fibers. Hybridization is one technique for enhancing the mechanical qualities of composites supplemented with natural fibers [50, 51].

3.5.1 FACTORS INFLUENCING THE MECHANICAL PROPERTIES

3.5.1.1 FIBER CONTENT

Fiber content is the percentage of fiber by weight or volume that is present in the composite. As the amount of fiber in the composite increases, so do qualities like tensile strength, flexural strength, and hardness. The maximum values of these characteristics were discovered at 40–50% of the fibers' weight fraction.

3.5.1.2 CHEMICAL TREATMENT

The tensile strength and flexural strength of the composites are enhanced by chemical or surface treatment of the fibers prior to composite manufacture. The qualities are improved by the surface treatment, which removes contaminants from fiber surfaces and strengthens the bond between the fiber and matrix. However, because the strength of the fiber after treatment decreases, the impact strength of the composites also decreases. Alkalization, silane treatment, acylation, and benzolization are a few examples of several chemical treatment methods.

3.5.1.3 MATRIX FIBER BONDING

The degree of bonding between the matrix and fiber greatly affects the characteristics of composites. Higher-quality composites with better matrix-to-fiber bonding have superior characteristics. Bonding between the matrix and fiber has a great impact on the properties of composites [32].

3.5.1.4 FIBER FORM

Long, short, and powdered fibers all fall under the category of fiber shape. Because longer fibers can absorb more energy, their impact strength grows as they get longer. Regarding the length of the fiber, the tensile and flexural strengths remain constant. Better tensile and flexural strength is provided by fibers that are the right size.

3.5.1.5 WATER ABSORPTION

The fibers' tendency to absorb water [53] weakens the bonds that hold the matrix and fibers together. Moreover, it causes the composites' characteristics to decline [54–57]. Hence, places with less water intrusion can use fiber-reinforced composites. Chemical processing can lessen this fiber behavior. The mechanism of water absorption is depicted in Figure 3.6.

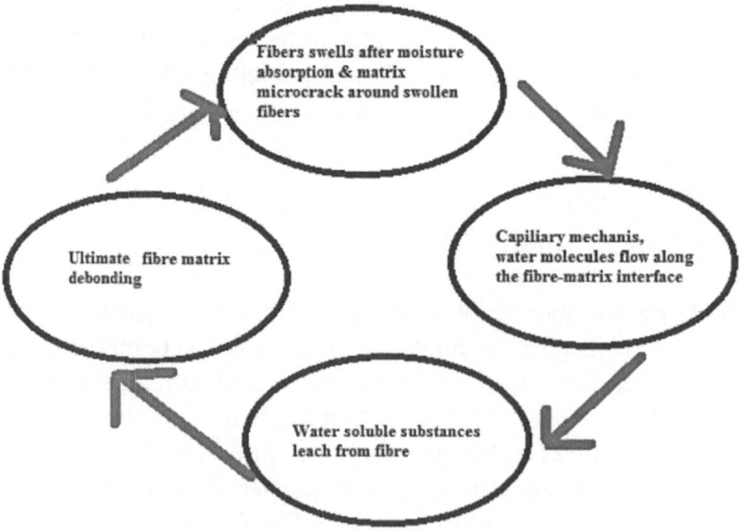

FIGURE 3.6 Mechanism of moisture absorption in natural fibers.

3.5.1.6 STACKING SEQUENCE

Several fibers are employed in some composites. The characteristics are also influenced by the way the fibers are arranged. Composites' characteristics can be enhanced by selecting the optimal fiber stacking order [58–61] (Figure 3.7).

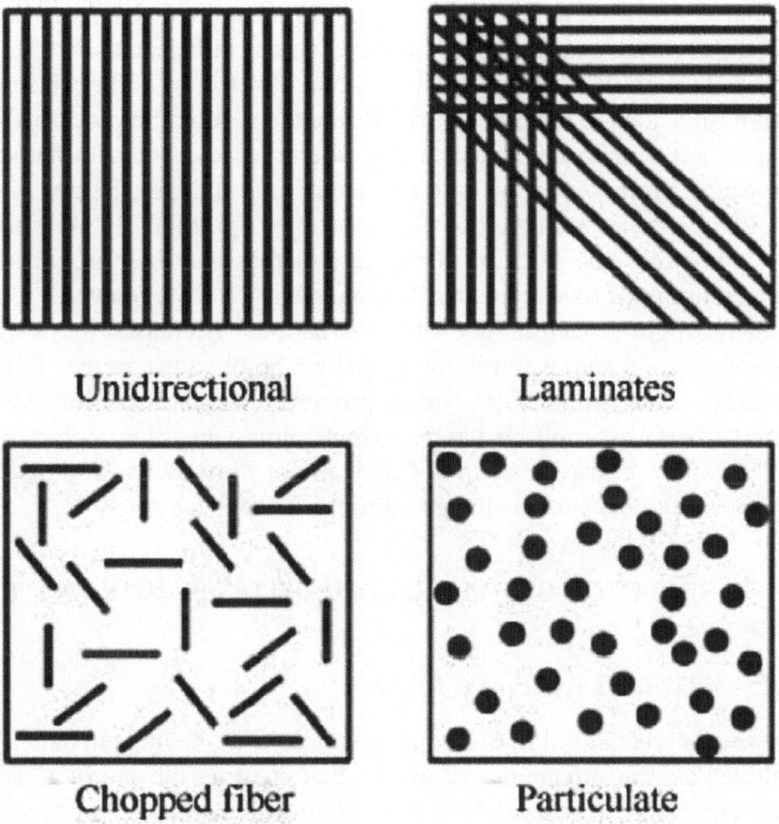

FIGURE 3.7 Various geometry of the fiber.

3.5.1.7 HYBRIDIZATION

Hybridization is the reinforcement of both natural and synthetic fibers in a matrix. Many fibers are used in hybrid composites. Hybridization results in a decrease in the use of synthetic fibers and an improvement in the proven properties of the composites. Compared to natural fiber reinforced composites, hybrid composites have more favorable features.

One of the major problems with natural fibers is their moisture absorption. Water is absorbed in fiber-reinforced polymer composites through diffusion, capillary action, and water molecule transfer. The diffusion mechanism occurs between the tiny gaps in the polymer chains. The natural fiber experiences differential swelling as a result of water diffusion at the fiber-matrix interface, mostly because of its hydrophilic makeup. Because of this, fiber swelling may cause strain at the interface, which could lead to matrix deterioration and micro-cracking and worsen water absorption. If the reinforcement was not fully soaked with the matrix during the production process, capillary transport happens in the spaces at the fiber-matrix contact. In natural fiber composites, the movement of water molecules through tiny fissures that might form in the matrix as a result of fiber swelling is particularly important. The ability of natural fibers to absorb moisture increases as their proportion does. Using compounds like silane, NaOH, and occasionally even water, chemical treatment might reduce it. The surface treatment decreased the weight and diameter of the fibers. Natural fibers can be employed in a range of industrial applications due to their mechanical properties and accessibility of supply. Natural fibers' hydrophilic nature, which diminishes the contact between the fiber and the matrix, is one of their major drawbacks. The hydrophilic properties of fibers can be changed through surface treatment.

3.6 FABRICATION OF HYBRID COMPOSITES BY USING NATURAL FIBER

3.6.1 MATERIALS AND METHODS

Epoxy resin and natural fibers from pineapple, flax, hemp, and bananas were employed as reinforcement materials when creating the samples. Table 3.1 lists the mechanical characteristics and chemical makeup of various fibers. A common type of leaf fiber found in nature is the fiber found in bananas and pineapples. In India, these are commonly accessible. A form of bast fiber, hemp and flax fibers are taken from the stalks of the plants. Used as a matrix material was epoxy resin and hardener.

3.6.1.1 PREPARATION OF FIBERS

One of the key components of the process is the selection and extraction of fiber. All of the recovered fibers must be thoroughly cleansed with water before being dried to remove the water. The fibers must then be slightly separated using a careful hand. The use of a mechanical combing procedure on the fiber laminates enabled the hand retting process to separate individual fiber strands. Following their separation, the fibers were repeatedly checked for filaments using a cotton checking tool. The remaining moisture was then removed by allowing the fibers to dry at room temperature. The weight and length of each fiber is then measured in accordance with the necessary measurements. Banana, flax, hemp, and other plant fibers are employed as reinforcement materials for the fabrication of samples.

3.6.1.2 PREPARATION OF MATRIX MATERIAL

For matrix material, epoxy resin or other types of materials are utilized in combination with hardeners. The weight ratios of both the resin and the hardener should be taken into account to obtain the correct resin solution. A plastic stirrer should be used to stir the epoxy resin and hardener for two to three minutes at room temperature until the mixture has a consistent hue. The mixture should then be swirled for an additional 30 seconds to scrape the container's bottom and sides. The hardener and epoxy should be well mixed before the resin solution is applied to the fibers.

3.6.1.3 WEIGHT FRACTION OF REINFORCEMENT AND MATRIX MATERIALS

The weight fraction of the constituents used for the fabrication process was considered based on the hybridization process.

3.6.1.4 COMPOSITE FABRICATION PROCEDURE

The hand layup technique is used during the fabrication process. The following processes are involved in the creation of coir fiber: First, the coir fibers should be thoroughly cleaned and dried in the shade. Next, the coir mold needs to be cleaned and kept free of dust before production can begin. After that, a coating of polyvinyl alcohol must be applied to the surface of the mold so that the laminate can be easily removed after manufacturing. The next step is to combine the epoxy resins (Araldite LY554) and hardener (amino hydrocarbon) in a 10:1 ratio. Then, lay the coir threads on top of the resin that has been applied to the mold with a brush. When fibers overlap, care must be taken to prevent variations in the finished composite's thickness. A roller can be used to securely place the coir fibers over the prior layer of glue and shuns.

Using a brush, apply two coats of resin to the coir layer. Cover the resin layer with the same number of coir fibers again, this time in a vertical direction. This will increase the overall strength of the composite, as a change in orientation can protect the material from the spread of applied stresses. Allow the lamination to rest for an hour. To prevent air bubbles from forming on the composite's surface, apply pressure.

Next, it is important to make sure the resin has relaxed before meticulously removing it from the mold and cutting it to the appropriate measurements. All samples must be processed using the same method. The same process should be used to create three samples with various oriented fibers, as shown in Figure 3.8.

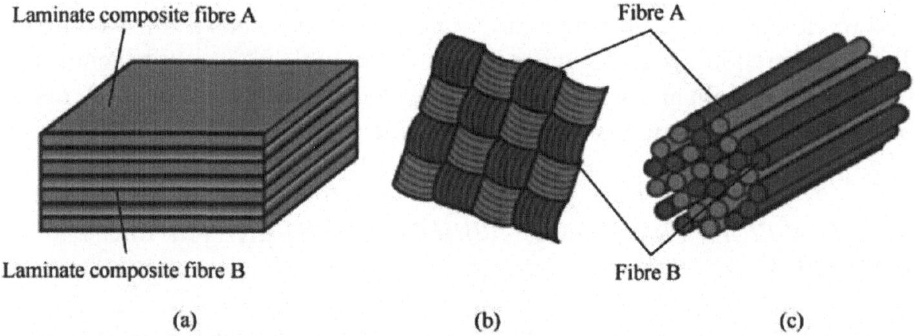

FIGURE 3.8 The three main hybrid configurations in fabrication are shown: (a) hybrid composite interlayer or layer-by-layer; (b) hybrid composite interlayer or yarn-by-yarn; and (c) hybrid composite intrayarn or fiber-by-fiber.

Source: Reprinted with permission from Ref. [65]. Copyright © 2014 Elsevier Ltd.

3.7 TESTING OF HYBRID COMPOSITES

3.7.1 TENSILE TEST

One of the most popular tests used by researchers to determine tensile qualities is the tensile test. Tensile tests must be conducted using the test specimens that the American Society for Testing and Materials (ASTM) has prescribed. The tensile testing equipment used in this study is from Associated Scientific Engineering Works, a member of the FIE group of India, and it has a gear rotation speed range of 1.25, 1.5, and 2.5 mm/min. It can test materials with a maximum weight of 5 tons. Repeated tests can be performed on specimens produced from three samples made of coir, hair, and a coir-hair hybrid. By clamping the specimen in the UTM at the ends, tensile tension is applied [60].

3.7.2 COMPRESSIVE TEST

A standard laminate piece that has been cut in accordance with ASTM standards is required for the tensile test. The test is carried out under controlled conditions in an Indian FIE group. Experiments are repeated using samples from three materials: coir, hair, and a coir-hair hybrid. The specimen is clamped in the UTM to apply compression force at the ends. Each sample reaches the intended compressive breaking load. Additionally, it is possible to depict the difference in compressive stress-strain.

3.7.3 FLEXURAL STRENGTH TEST

To determine the mechanical characteristics of the composite materials, flexural tests are conducted. The maximum stress and strain resulting from the application of an external load are calculated using the test results.

3.7.4 FLEXURAL MODULUS TEST

The capacity of a composite material to deform is known as its flexural modulus. It is determined using the stress and displacement curve's slope. It is also known as a tangent modulus and an elastic modulus [60].

3.7.5 IMPACT TEST

The material receives a lot of energy during the impact test. An ASTM-compliant Charpy Impact Machine can be used to conduct this impact test. The samples for this test must be cut to the prescribed measurements. The specimen yields to the powerful blow delivered by the hammer, and the force of the blow it is unable to endure is what causes the impact. It is necessary to repeat the tests on the three samples of specimens made of coir, hair, and a coir-hair hybrid.

3.7.6 *INTERLAMINAR SHEAR STRENGTH*

The polymer industry uses a variety of shear tests to determine the strength of the reinforcement to matrix materials. The failure shear that takes place in a composite material when a transverse load is applied is called the interlaminar shear strength.

3.7.7 *MOISTURE ABSORPTION TEST*

The Moisture Absorption Test, also known as the 7.7 test, is performed to determine how quickly the composite will absorb moisture (water) when exposed to a damp environment. The composite sample is placed in a container with distilled water to measure the rate of absorption. The samples are examined in accordance with ASTM D-570.

The percentage weight gain is calculated using the formula:

Percentage Weight Gain = (Final weight – Initial weight/Initial weight) × 100

3.7.8 *HARDNESS TEST*

The indentation rigidity of materials based on polymers is estimated by the hardness test. According to ASTM standards, the test is conducted. Using a hardened steel indenter with a specific force and geometry, the specimen must be indented [15, 61].

3.8 FUNDAMENTAL PROPERTIES

The properties like density, tensile strength, modulus, percentage of elongation, and moisture content are shown in Table 3.2.

- Density;
- Tensile strength;
- Modulus;
- Percentage of elongation;
- Moisture content.

TABLE 3.2 Natural Fiber's Properties

IBER	Density (G/CM3)	Tensile Strength (MPA)	Young's Modulus (GPA)	Elongation at Break (%)	Moisture Content (wt.%)
Flax	1.5	345–1,100	27.6	2.7–3.2	10.0
Hemp	1.47	690	30–60	1.6	10.8
Jute	1.3–1.4	393–773	13.0–26.5	1.2–1.5	12.6
Kenaf	1.3	930	53	1.6	–
Ramie	1.5	400–938	61.4–128	1.2–3.8	16
Abaca	1.5	400	12.0	3.0–10.0	15
Pineapple	1.3	413–1,627	34.5–82.7	1.6	11.8
Sisal	1.4	468–640	9.4–22.0	3.0–7.0	11.0
Coir	1.1	131–175	4.0–6.0	15–40	8.0
Cotton	1.5–1.6	287–800	5.5–12.6	7.0–8.0	19
E-glass	2.55	2,400	73	3	–

Source: Fictitious data, for illustration purposes only.

3.9 ADVANTAGES OF NATURAL FIBER COMPOSITES

3.9.1 HIGH STRENGTH

Metals with high strength in all directions, such as steel or aluminum, are commonly used in the construction industry. Composite materials must be designed in a specific way to be made stronger than conventional materials.

3.9.2 DURABILITY

Composites have a long lifespan and require little maintenance when used in construction. Many composites have a lifespan of about 50 years. However, we do not know how long a composite structure can demonstrate its durability, nor do we know when the original composites have reached the end of their useful lives.

3.9.3 LOW THERMAL CONDUCTIVITY

Composites made of natural fibers are effective insulators. They are used for doors, windows, and panels to protect structures from the elements because composites made of natural fibers do not easily transfer heat or cold.

3.9.4 STRENGTH RELATED TO WEIGHT

The strength-to-weight ratio is an attribute of composite materials that provides information about their strength, weight, and lightness. Composite materials are not only strong but also lightweight, which makes them suitable for use in the automobile and aircraft industries.

3.9.5 LIGHTWEIGHT

Composites are used in aeronautical and automotive applications because they are lighter than other materials such as metals and wood. Modern airplanes, such as the Boeing 787 Dreamliner and Cockpit, are being constructed using composite materials because fuel efficiency is a growing priority, which can only be achieved by using lightweight composite materials.

3.9.6 CORROSION RESISTANCE AND HIGH-IMPACT STRENGTH

Composites offer strong impact resistance and good corrosion resistance. They have the ability to absorb anything and withstand weather-related damage. The chemical resistance of composites is good, and they can withstand severe climates. Due to their ability to absorb abrupt forces, composite materials are used to create bulletproof jackets and panels, as well as to protect buildings, military vehicles, and aircraft from explosions.

3.9.7 DIMENSIONAL STABILITY

Composites maintain constant conditions of stability. Weather or temperature have little effect on their size or shape. On the other hand, when the temperature changes, wood contracts and expands. As a result, composites are utilized to manufacture aircraft wings because of their exceptional ability to adapt to any circumstance and maintain their shape and size regardless of how hot or how cold it is outside.

3.9.8 NONCONDUCTIVE

Composites possess a nonconductive quality. It implies that composites are not electrically conductive. Circuit boards and electrical utility poles are both made of composite materials due to this characteristic. Since composite materials must occasionally have electrical conductivity, it is simple to change their characteristics to become conductive materials.

3.9.9 NONMAGNETIC

The fact that composites lack magnetic qualities makes them the sole type of material that can be utilized to create houses, tables, enclosures for electrical equipment, and reinforced concrete for hospital walls and flooring.

3.9.10 RADAR TRANSPARENT

Radar equipment is made of composites, and composite materials are utilized in radar to transmit signals in the air or on the ground. The U.S. Air Force's B-2 stealth bomber, which is employed for signal transmission, is made in large part from composite materials.

3.9.11 BIODEGRADABILITY

Natural fiber reinforcing in polymers produces high-strength composites, which also offer additional or increased biodegradability, low cost, light weight, and superior mechanical structure features. Natural fibers begin to break down at 240°C, but other components of fiber, such as cellulose, lignin, and others, begin to break down at different temperatures. For instance, lignin begins to break down at 200°C, but other components will also begin to break down at temperatures higher than this. As the structural components of fibers, such as lignin and hemicelluloses, determine the thermal stability of the fibers, it is possible to increase the concentration of these components or totally remove them. Chemical treatments can assist in achieving this. When natural fibers degrade, two crucial factors should be taken into account: the development of fibers and materials that offer services. While synthetic fibers hurt the environment owing to pollution brought on by degradation, natural fibers have a short lifespan and provide the least amount of environmental harm. The concentrations or composition of lignin, hemicelluloses, and cellulose affect the thermal breakdown characteristics of lignocellulosic materials. At exactly 1,500 days of burial, more than 50% of the weight of jute composite is lost.

3.10 NOVEL APPLICATION OF HYBRID POLYMER COMPOSITE

The number of uses being discovered for fiber reinforced polymers is also steadily rising. Some of these systems, however, are only effective in highly specialized settings where drawbacks like high cost and brittle fracture behavior are viewed as secondary to benefits like low density, high rigidity, and high strength, as well as in less demanding applications where weight is not as important.

3.10.1 NANOTECHNOLOGY

In the realm of nanotechnology, composite materials have a wide variety of uses. The nanocomposite is an innovative design that contains fillers isolated in the resin at the nanoscale. With the addition of only a small amount of fillers, the resin's flame retardance and rigidity are considerably improved due to the diffusion of extremely small fillers. Polymer nanofibers are utilized in medical applications to treat burns and wounds on human skin, as well as to construct hemostatic methods and equipment with unique and unusual properties.

3.10.2 WIND POWER GENERATION

Wind power engineering is a top priority for energy production due to its ability to conserve resources and be environmentally safe. Basic power element blades have a significant impact on how much energy is used.

3.10.3 HYBRID THERMOPLASTIC APPLICATION

Advanced thermoplastic composites have long had the potential to produce lightweight structural components in large quantities. The Army created the helmet design as part of the Future Force Warrior (FFW) program. The composite material used to create these helmets, known as PASGT helmets, consists of aramid fabric in a thermoset matrix. Also, the building must be sturdy enough to resist daily wear from a soldier's activities and offer better ballistic protection.

3.10.4 AEROSPACE APPLICATIONS

Due to their improved strength, durability, corrosion resistance, resilience to fatigue, and damage tolerance properties, fiber-reinforced composites have become an increasingly appealing alternative to traditional metals for many aviation components in contemporary civil aircraft built to meet various power and safety requirements. The best chosen materials are glass and carbon-reinforced hybrid composites because of forward-thinking technology that has gone beyond design and application.

3.10.5 SMART MEMORY HYBRID COMPOSITES

Increasing demands for the use of materials in engineering practical applications presuppose the creation of these materials, which are therefore referred to as adaptive, multifunctional, smart, or intelligent materials. The SMA elements improve the physical qualities of the matrix materials.

3.10.6 CHEMICAL INDUSTRY

Due to its many benefits, including its light weight, improved fire resistance, and chemical resistance, composite materials are one of the most often utilized materials in the chemical industry. In the chemical sector, composite grips are frequently used to create structural supports, storage tanks, exhaust stacks and blowers, columns, pumps, reactors, and other items for acidic and alkaline environments.

3.10.7 ORTHOPEDIC AIDS PROSTHETICS AND ORTHOTICS

By providing them with artificial supports, you can assist those who have a disability or were born with physical flaws. A prosthesis is a synthetic replacement for a missing body part. The artificial body parts that replace missing arms and legs are the ones that are most frequently thought of as prostheses, although prostheses can also include replacements for bones, arteries, heart valves, artificial eyes, and teeth.

3.10.8 AUTOMOBILE/TRANSPORTATION SECTOR

Despite the potential benefits of lighter weight and high performance resulting from corrosion resistance, leading composites are not usually applied in the automotive industry. There must be some action taken on a worldwide level to make high-performance composite materials attractive for extensive use in automobiles and other areas [32].

3.10.9 BRIDGE CONSTRUCTION

In many countries, research and development for totally hybrid FRP structures in civil engineering [42, 48] is in progress. In modern structural elements, hybrid fiber-reinforced plastics (GFRP and CFRP) are used extensively. Nevertheless, HFRP solutions are used for bridges due to the considerable corrosive atmosphere. The bridges' extruded composite components convey serious mechanical properties at a lower weight than steel. For instance, the West Mill Bridge has a span of 10 m and a width of 6.8 meters. The total weight of the construction is 37 tons, with a load-carrying capacity of 46 tons and an axle load of 13.5 tons (Figure 3.9).

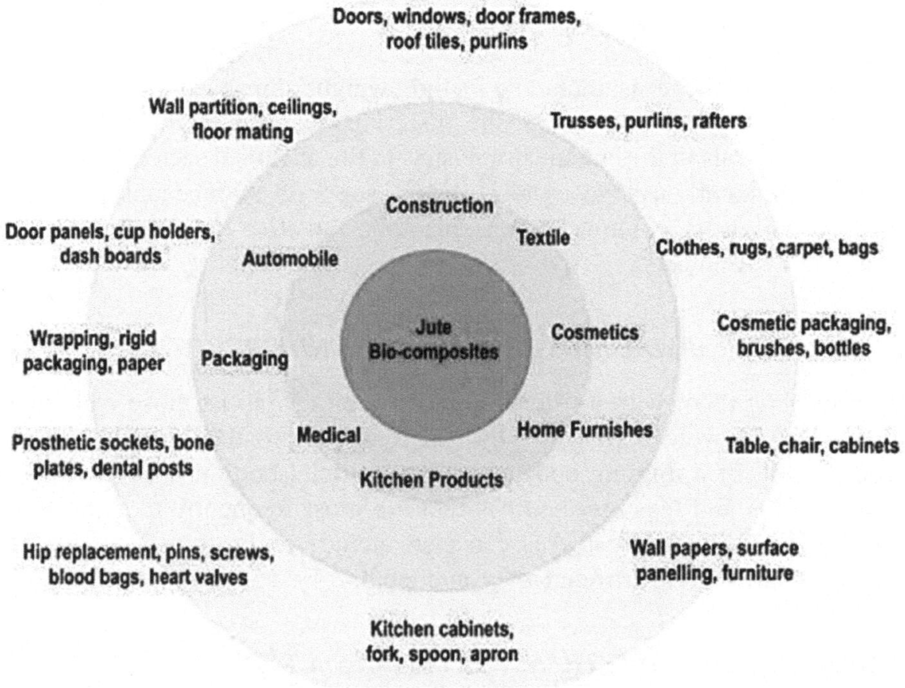

FIGURE 3.9 Bio- and hybrid-composites and their applications.

Source: Reprinted with permission from Ref. [64]. Copyright © 2019 by the authors. Licensee MDPI, Basel, Switzerland. http://creativecommons.org/licenses/by/4.0/

3.11 MERITS AND DEMERITS

➢ **Merits:**

1. Fuel savings result from the increased performance at a given weight. Composite materials can attain fantastic strength-to-weight and stiffness-to-weight ratios. Strength divided by density and stiffness (modulus) divided by density are two common ways to represent this. They are referred to as having "specific" strength and "specific" modulus properties.
2. Part count has also been reduced.
3. Smooth aerodynamic profiles are simpler to develop and easier to reduce drag. It is possible to produce intricate double-curvature pieces with a smooth surface finish in a single manufacturing step.
4. The part's laminate patterns and ply build-up can be customized to provide the necessary mechanical qualities in different directions.
5. Composites provide excellent protection against corrosion, chemical attack, and outdoor weathering. However, some chemicals (such as paint stripper) can damage composites, so new paint and stripper formulations are being developed to address this issue. Thermoplastics are also not very resistant to some solvents. Verify the type's data sheets.
6. Production cost is reduced. Composites may be made by a wide range of processes.

➢ **Demerits:**

1. Composites were more easily damaged than wrought metals because they are more fragile. Cast metals are frequently fragile as well.
2. Repair causes new issues because materials must be transported and stored in a refrigerator and have a finite shelf life. In many instances, hot curing is important and calls for specialized tools. Both hot and cold cure require time. When the final rivet is set, the project is not complete.
3. It can be difficult to remove rivets that have been utilized and must be removed without causing more harm.
4. Before repairing composites, contaminants must be completely removed.

5. Tooling under pressure is necessary for repair at these original cure temperatures.
6. Prior to repair, composites must be dried since some fibers and all resin matrices absorb moisture.

3.12 OPPORTUNITIES, CHALLENGES, AND FUTURE TRENDS

To expand the use of hybrid composites in engineering, several obstacles [9] must be addressed. To meet these obstacles, one needs design, research, product development, and business development abilities. It is essential to address the following concerns in this endeavor: comprehending the science behind hybrids' initial processing, particularly the variables that determine their microstructural integrity; improving fracture toughness and ductility for damage-tolerant qualities in hybrid composites; making efforts to create inexpensive, high-quality reinforcements using industrial waste and byproducts; prioritizing the creation of hybrids based on uncommon fibers and matrices; and categorizing different grades of hybrids according to their property profiles. Additionally, there is an urgent need to develop simple, economical, and portable non-destructive kits to quantify undesirable defects in hybrid composites [16, 27, 31, 65].

3.13 CONCLUSION

Natural fiber is a renewable, environmentally beneficial, and biodegradable raw material. Both its mechanical and thermal insulation qualities are strong in natural fiber. Yet, the loading placed on the natural fiber and its Young's modulus determine its strength. It also depends on the fiber weight ratio, the growing process, the plant's method for extracting fiber, the timing of harvest, the manufacturing process, and the techniques used to make natural fibers with polymeric matrices. In the fields of industry, engineering, and technology, natural fiber composites are a novel, versatile material created by combining two different types of materials. In this chapter, we reviewed natural fibers, their various varieties, their physical and mechanical characteristics, their benefits, and their disadvantages in comparison to other materials. In a nutshell, natural fiber utilization in industrial applications is rising quickly. The limitations of natural fiber, such as stiffness of composites, moisture effect, alkaline solution, fatigue, creep, and physical degradation, moisture absorption toughness, and reduced long-term stability for outdoor application, have already been the subject of extensive research, but more needs to be done in this area. Many academics are still very interested in studying natural fibers and composites made of natural fibers.

Since natural fiber-based composite materials are durable, renewable, and partially or entirely recyclable, research on them has increased dramatically during the past few years. Despite all of the aforementioned problems, natural fibers are nevertheless used in a number of industries to create innovative goods.

KEYWORDS

- **biodegradable raw material**
- **fiber utilization**
- **industrial applications**
- **moisture effect**
- **natural fibers**
- **Young's modulus**

REFERENCES

1. Haris, N. I. N., Ilyas, R. A., Hassan, M. Z., Sapuan, S. M., Afdzaluddin, A., Jamaludin, K. R., Zaki, S. A., & Ramlie, F., (2021). Dynamic mechanical properties and thermal properties of longitudinal basalt/woven glass fiber reinforced unsaturated polyester hybrid composites. *Polymers*, *13*, 3343.
2. Sapuan, S. M., Aulia, H. S., Ilyas, R. A., Atiqah, A., Dele-Afolabi, T. T., Nurazzi, M. N., Supian, A. B. M., & Atikah, M. S. N., (2020). Mechanical properties of longitudinal basalt/woven-glass-fiber-reinforced unsaturated polyester-resin hybrid composites. *Polymers*, *12*, 2211.
3. Sanjay, M. R., Arpitha, G. R., Naik, L. L., Gopalakrishna, K., & Yogesha, B., (2016). Applications of natural fibers and its composites: An overview. *Nat. Resour.*, *7*, 108–114.
4. Sanjay, M. R., Arpitha, G. R., Senthamaraikannan, P., Kathiresan, M., Saibalaji, M. A., & Yogesha, B., (2019). The hybrid effect of jute/kenaf/E-glass woven fabric epoxy composites for medium load applications: Impact, inter-laminar strength, and failure surface characterization. *J. Nat. Fibers*, *16*, 600–612.
5. Madhu, P., Sanjay, M. R., Senthamaraikannan, P., Pradeep, S., Saravanakumar, S. S., & Yogesha, B., (2019). A review on synthesis and characterization of commercially available natural fibers: Part II. *J. Nat. Fibers*, *16*, 25–36.
6. Ghelli, D., & Minak, G., (2011). Low velocity impact and compression after impact tests on thin carbon/epoxy laminates. *Compos. Part. B Eng.*, *42*, 2067–2079.

7. Boria, S., Pavlovic, A., Fragassa, C., & Santulli, C., (2016). Modeling of falling weight impact behavior of hybrid basalt/flax vinylester composites. *Procedia Eng.*, *167*, 223–230.
8. Wambua, P., Ivens, J., & Verpoest, I., (2003). Natural fibers: Can they replace glass in fiber reinforced plastics *Composites Science and Technology*, *63*, 1259–1264.
9. Affan, U. M., & Anas, M., (2018). Study of natural fiber reinforced composites. *IOP Conference Series: Materials Science and Engineering*, *404*, 012–048. https://doi.org/10.1088/1757-899X/404/1/012048.
10. Balaji, A., Karthikeyan, B., & Sundar, R. C., (2015). bagasse fiber – the future biocomposite material: A review. *International Journal of Chem Tech Research*, *7*(1), 223–233.
11. Bongarde, U. S., & Shinde, V. D., (2014). Review on natural fiber reinforcement polymer composites, *International Journal of Engineering Science and Innovative Technology (IJESIT), 3*(2), 431–436.
12. Vijay, K. T., Manju, K. T., & Raju, K. G., (2014). Review: Raw natural fiber-based polymer composites. *International Journal of Polymer Analysis and Characterization*, *19*(3), 256–271. https://doi.org/10.1080/1023666X.2014.880016.
13. Staiger, M. P., & Tucker, N., (2008). Natural-fiber composites in structural applications. In: Pickering, K. L., (ed.), *Properties and Performance of Natural Fiber Composites* (pp. 269–300). Cambridge: Woodhead Publishing Limited and CRC Press LLC. https://doi.org/10.1533/9781845694593.2.269.
14. Ramesh, M., Palanikumar, K., & Reddy, K. H., (2017). Plant fiber-based bio-composites: Sustainable and renewable green materials. *Renew. Sustain. Energy Rev.*, *79*, 558–584.
15. Thyavihalli, G. Y. G., Mavinkere, R. S., Parameswaranpillai, J., & Siengchin, S., (2019). Natural fibers as sustainable and renewable resource for development of eco-friendly composites: A comprehensive review. *Front. Mater.*, *6*, 1–14.
16. Keller, T., (2003). Use of fiber reinforced polymers in bridge construction. *International Association for Bridge and Structural Engineering*. Zurich.
17. Saira, T., Munawar, M. A., & Khan, S., (2007). Natural fiber- reinforced polymer composites. *Proceedings of the Pakistan Academy of Sciences, 44*(2), 129–144.
18. Faruk, O., Bledzki, A. K., Fink, H. P., et al., (2012). Biocomposites reinforced with natural fibers: 2000–2010. *Progress in Polymer Science*, *37*(11), 1552–1596. https://doi.org/10.1016/j.progpolymsci.2012.04.003.
19. Nguong, C. W., Lee, S. N. B., & Sujan, D., (2013). A review on natural fiber reinforced polymer composites. *International Journal of Materials and Metallurgical Engineering*, *7*(1), 52–59.
20. Zini, E., & Scandola, M., (2011). Green composites: An overview. *Polymer Composites, 32*(12), 1905–1915. https://doi.org/10.1002/pc.21224.
21. Ticoalu, A., Aravinthan, T., & Cardona, T., (2010). A review of current development in natural fiber composites for structural and infrastructure applications. *Southern Region Engineering Conference (SREC'10)* (pp. 113–117). Toowoomba, Australia.
22. Abilash, N., & Sivapragash, M., (2013). Environmental benefits of eco-friendly natural fiber reinforced polymeric composite materials. *International Journal of Application or Innovation in Engineering and Management (IJAIEM)*, *2*(1), 53–59.
23. Akil, H. M., Omar, M. F., Mazuki, A. A. M., et al., (2011). Kenaf fiber reinforced composites: A review. *Material and Design, 32*(8, 9), 4107–4121. https://doi.org/10.1016/j.matdes.2011.04.008.

24. Sailesha, A., Arunkumar, R., & Saravanan, S., (2018). Mechanical properties and wear properties of kenaf – aloe vera – jute fiber reinforced natural fiber composites. *Materials Today: Proceedings*, *5*, 7184–7190.
25. Ahmad, F., Choi, H. S., & Park, M. K., (2015). A review: Natural fiber composites selection in view of mechanical, light weight, and economic properties. *Macromol. Mater. Eng.*, *300*(1), 10e24.
26. Pickering, K. L., Efendy, M., & Le, T., (2016). A review of recent developments in natural fiber composites and their mechanical performance. Compos. *A Appl. Sci. Manuf.*, *98*–112.
27. Reddya, M. I., Varmaa, U. R. P., Kumara, I. A., Manikantha, V., & Raju, P. V. K., (2018). Comparative evaluation on mechanical properties of jute, pineapple leaf fiber and glass fiber reinforced composites with polyester and epoxy resin matrices. *Materials Today: Proceedings, 5*, 5649–5654.
28. Reddy, M. I., Kumar, M. A., & Raju, C. R. B., (2018). Tensile and flexural properties of jute, pineapple leaf and glass fiber reinforced polymer matrix hybrid composites. *Materials Today: Proceedings*, *5*, 458–462.
29. Gunti, R., & Atluri, R. P., (2013). Study on effect of chemical treatments and concentration of jute on tensile properties of long and continuous twisted jute/polypropylene composite. *Advanced Materials Manufacturing and Characterization*, *3*(1), 395–398. https://doi.org/10.11127/ijammc.2013.02.073.
30. Muhammad, A. A., Mohammed, Z., Muhammad, T. M., Ramesh, K., & Ali, B., (2019). *Fibers*, *7*, 77. https://doi:10.3390/fib7090077.
31. Safri, et al., (2018). *Composites Part B, 133*, 112e121.
32. Reddy, M. I., Kumar, M. A., & Raju, C. R. B., (2018). Tensile and flexural properties of jute, pineapple leaf and glass fiber reinforced polymer matrix hybrid composites. *Materials Today: Proceedings*, *5*, 458–462.
33. Babatunde, O. E., Yatim, J. M., Ishak, M. Y., et al., (2015). Potentials of kenaf fiber in bio-composite production: A review. *Jurnal Teknologi (Sciences & Engineering)*, *77*(12), 23–30. https://doi.org/10.11113/jt.v77.6304.
34. Akash, Venkatesha, G. N S., & Sreenivas, R. K. V., (2018). An experimental study on sisal/hemp fiber reinforced hybrid composites, an experimental study on sisal/hemp fiber reinforced hybrid composites. *Materials Today: Proceedings*, *5*, 7383–7387.
35. Yousif, B. F., & Ku, H., (2012). Suitability of using coir fiber/polymeric composite for the design of liquid storage tanks. *Material & Design*, *36*, 847–853. https://doi.org/10.1016/j.matdes.2011.01.063.
36. Gu, H., (2009). Tensile behaviors of the coir fiber and related composites after NaOH treatment. *Materials and Design*, *30*, 3931–3934.
37. Nadir, A., Songklod, J., Vallayuth, F., et al., (2011). Coir fiber reinforced polypropylene composite panel for automotive interior applications. *Fibers and Polymer*, *12*(7), 919–926.https://doi.org/10.1007/s12221-011-0919-1.
38. Verma, D., Gope, P. C., Shandilya, A., et al., (2013). Coir fiber reinforcement and application in polymer composites: A review. *Journal of Material Environment Science, 4*(2), 263–276.
39. Charoenvaisarocha, A. C., Jongjit, H., & Joseph, K., (2009). Materials and mechanical properties of pretreated coir-based green composites. *Composites: Part B*, *40*, 633–637. https://doi.org/10.1016/j.compositesb.2009.04.009.

40. Siakeng, R., Jawaid, M., Asim, M., Saba, N., Sanjay, M. R., Siengchin, S., & Fouad, H., (2020). Alkali treated coir/pineapple leaf fibers reinforced PLA hybrid composites: Evaluation of mechanical, morphological, thermal and physical properties. *Xpress Polym. Lett.*, *14*, 717–730.
41. Kumre, A., Rana, R. S., & Purohit, R., (2017). A Review on mechanical property of sisal glass fiber reinforced polymer composites. *Materials Today: Proceedings*, *4*, 3466–3476.
42. Ali, M., (2012). Natural fibers as construction materials. *Journal of Civil Engineering and Construction Technology*, *3*(3), 80–89. https://doi.org/10.5897/JCECT11.100.
43. Thwe, M. M., & Liao, K., (2000). Characterization of bamboo-glass fiber reinforced polymer matrix hybrid composite. *Journal Of Materials Science Letters*, *19*, 1873–1876. https://doi.org/10.1023/A:1006731531661.
44. Jiang, L., Huang, J., Qian, J., et al., (2008). Study of poly(3- hydroxybutyrate-co-3-hydroxyvalerate) (PHBV)/bamboo pulp fiber composites: Effects of nucleation agent and compatibilizer. *Journal of Polymers and the Environment*, *16*(2), 83–93. https://doi.org/10.1007/s10924-008-0086-7.
45. Okubo, K., Fujii, T., & Yamamoto, Y., (2004). Development of bamboo-based polymer composites and their mechanical properties. *Composites Part A: Applied Science and Manufacturing*, *35*(3), 377–383. https://doi.org/10.1016/j.compositesa.2003.09.017.
46. Sinha, A. K., Narang, H. K., & Bhattacharya, S., (2017). Effect of alkali treatment on surface morphology of abaca fiber *Materials Today: Proceedings*, *4*, 8993–8996.
47. Rashad, M., Pan, F., Yu, Z., et al., (2015). Investigation on microstructural, mechanical and electrochemical properties of aluminum composites reinforced with graphene nanoplatelets. *Prog. Nat. Sci. Mater.*, *25*,460–70.
48. Uddin, N. E., (2013). *Developments in Fiber-Reinforced Polymer (FRP) Composites for Civil Engineering*. Elsevier. https://doi.org/10.1533/9780857098955.
49. Swolfs, Y., Gorbatikh, L., & Verpoest, I., (2014). **Fiber hybridization in polymer composites: A review.** *Compos Part A Appl. Sci. Manuf*, *67*, 181–200.
50. Kumar, A. K., Abishek, T., Jegatheesh, R. S., & Gokul, K. A., (2016). A review on mechanical properties of natural fiber reinforced hybrid polymer composites. *International Journal of Advanced Scientific and Technical Research*, *21*.
51. Noorunnisa, K. P., Abdul, K. H. P. S., Jawaid, M., Ramachandra, R. G., Surya, N. C., & Venkata, N. S., (2010). Sisal/carbon fiber reinforced hybrid composites: Tensile, flexural and chemical resistance properties. *J. Polym. Environ.*, *18*, 727–733.
52. Kim, D. H., Kim, H. G., & Kim, H. S., (2015). Design optimization and manufacture of hybridglass/carbonfiber reinforced composite bumper beam for automobile vehicle. *Compos. Struct.*, *131*, 742e52.
53. Rostamiyan, Y., Fereidoon, A., Mashhadzadeh, A. H., Ashtiyani, M. R., & Salmankhani, A., (2015). Using response surface methodology for modeling and optimizing tensile and impact strength properties of fiber orientated quaternary hybrid nano composite. *Compos Part B Eng*, *69*,304e16.
54. Asim, M., Jawaid, M., Abdan, K., & Ishak, M. O. R., (2016). Effect of alkali and silane treatments on mechanical and fiber-matrix bond strength of kenaf and pineapple leaf fibers. *Journal of Bionic Engineering*, *13*, 426–435.
55. Ashik, K. P., Sharma, R. S., & Guptha, V. L. J., (2018). Investigation of moisture absorption and mechanical properties of natural /glass fiber reinforced polymer hybrid composites. *Materials Today: Proceedings*, *5*, 3000–3007.

56. Alomayri, T., Assaedi, H., Shaikh, F. U. A., & Low, I. M., (2014). Effect of water absorption on the mechanical properties of cotton fabric-reinforced geopolymer composites. *J. Asian Ceram. Soc.*, *2*, 223–230. doi: 10.1016/j.jascer.2014.05.005.
57. Paturel, A., & Dhakal, H. N., (2020). Influence of water absorption on the low velocity falling weight impact damage behavior of flax/glass reinforced vinyl ester hybrid composites. *Molecules*, *25*, 278. doi: 10.3390/*molecules*25020278.
58. Alamri, H., & Low, I. M., (2012). Mechanical properties and water absorption behavior of recycled cellulose fiber reinforced epoxy composites. *Polym. Test.*, *31*, 620–628. doi: 10.1016/*j.polymertesting*.2012.04.002.
59. Athijayamani, A., Thiruchitrambalam, M., Natarajan, U., et al., (2009). Effect of moisture absorption on the mechanical properties of randomly oriented natural fibers/ polyester hybrid composite. *Materials Science and Engineering: A*, *517*(1), 344–353. https://doi.org/10.1016/j.msea.2009.04.027.
60. Chadramohan, D., & Marimuthu, K., (2011). Tensile and hardness tests on natural fiber reinforced polymer composite material. *International Journal of Advanced Engineering Science and Technologies, 6*(1), 97–104.
61. Balasubramanian, K., (2019). A review on recent application of hybrid composites; *International Journal of Management, Technology and Engineering, IX*, I.
62. Oqla, A. L., & Sapuan, F. S., (2014). Natural fiber-reinforced polymer composites in industrial applications: Feasibility of date palm fibers for sustainable automotive industry. *J. Clean. Prod.,* 347–354.
63. Loganathan, M. T., Sultan, M. T. H., Gobalakrishnan, M. K. and Muthaiyah, G. (2018). Ballistic impact response of laminated hybrid composite materials 10. 10.1016/ B978-0-08-102292-4.00010-2.
64. Ashraf, M.A.; Zwawi, M.; Taqi Mehran, M.; Kanthasamy, R.; Bahadar, A. Jute Based Bio and Hybrid Composites and Their Applications. Fibers 2019, 7, 77. https://doi.org/10.3390/fib7090077.
65. Swolfs, Yentl & Gorbatikh, Larissa & Verpoest, I.. (2014). Fibre hybridisation in polymer composites: A review. Composites Part A: Applied Science and Manufacturing. 67. 181–200. 10.1016/j.compositesa.2014.08.027.

CHAPTER 4

Carbon Nanotubes for Sensor and Energy Applications

CINTIL JOSE CHIRAYIL,[1] ANNU JORLY,[1] ABIYA VARGHESE,[1] SNEHA JIPSON,[1] ANITTA MATHEW,[2] and SABU THOMAS[3]

[1]*Department of Chemistry, Newman College, Thodupuzha, Idukki, Kerala, India*

[2]*Department of Chemistry, Pavanatma College, Murickassery, Idukki, Kerala, India*

[3]*School of Chemical Sciences, Mahatma Gandhi University, Kottayam, Kerala, India*

ABSTRACT

Carbon nanotubes (CNTs) have attracted a lot of interest in the field of electrochemical sensing and in the field of energy because of their unique structural, electronic, and chemical properties, such as their tubular nanostructure, large specific surface, excellent conductivity, modifiable sidewall, high conductivity, and good biocompatibility. Carbon nanotubes (CNTs) can be used as polymer and catalyst additives, in autoelectron emission for cathode rays of lighting components, flat displays, gas discharge tubes in telecommunication networks, absorption, and screening of electromagnetic waves, energy conversion, lithium battery anodes, hydrogen storage, composite materials (fillers or coatings), nanoprobes, sensors, supercapacitors, and many other applications. This chapter focuses on the sensor applications of carbon nanotubes in detail.

Sustainable Environmental Engineering: Water Security, Energy Conservation, and Green Process.
Sonia Khanna, Ann Rose Abraham, Kian Hariri Asli, & A. K. Haghi (Eds.)

4.1 INTRODUCTION

With thousands of publications reported over the years, carbon nanotubes (CNTs) are one of the most commonly used nanomaterials. CNTs have several features that make them ideal for chemical and biological sensing because of their unique structure and nanoscale size. These characteristics include high surface area to volume and aspect ratios, superior electrical conductivity, chemical stability, and luminescent qualities. Electrochemical, photonic, and field-effect devices have all been used to apply CNTs to biosensors. CNTs have numerous advantages in the fabrication of electrochemical sensors, including – (i) large specific area that produces high sensitivity; (ii) a tubular nanostructure and chemical stability that allows the fabrication of ultrasensitive sensors using only one nanotube; (iii) good biocompatibility that is an excellent choice for the construction of electrochemical biosensors, particularly for the promotion of the electron transfer of redox proteins and enzymes; and (iv) the modifiability of the nanotubes [1–10].

4.2 HISTORY, STRUCTURE, AND PROPERTIES

Carbon, which came to Earth from red giant stars, is a unique element in the periodic table because it can bond with itself or other atoms without using a lot of energy. Carbon has long been recognized to exist in three allotropic forms (diamond, graphite, and amorphous carbon), all of which are essential for life. In the middle of the 20th century, significant advancements in carbon science led to unexpected and exciting discoveries. For example, hexagonal diamond was discovered as minuscule crystals coupled with diamond in the Canyon Diablo meteorite in 1967. The discovery of buckminsterfullerenes (C_{60}) later marked the beginning of a modern age in carbon science [11–15].

CNTs are classified as single-walled (SWNTs) or multi-walled (MWNTs) depending on the number of walls they have (MWNTs). SWNTs can grow up to 20 cm in length with diameters as small as a nanometer (or less). The side walls of these tubes are constructed of a hexagonal lattice of carbon atoms, similar to the atomic planes of graphene, and one half of a fullerene-like molecule is usually capped at both ends. SWNTs have the simplest form and can be seen as a single graphene sheet wrapped up. The structure of a nanotube can be simply specified by the chiral vector, which is given by the orientation of the tube axis with respect to the hexagonal lattice [16–19] as shown in Figure 4.1.

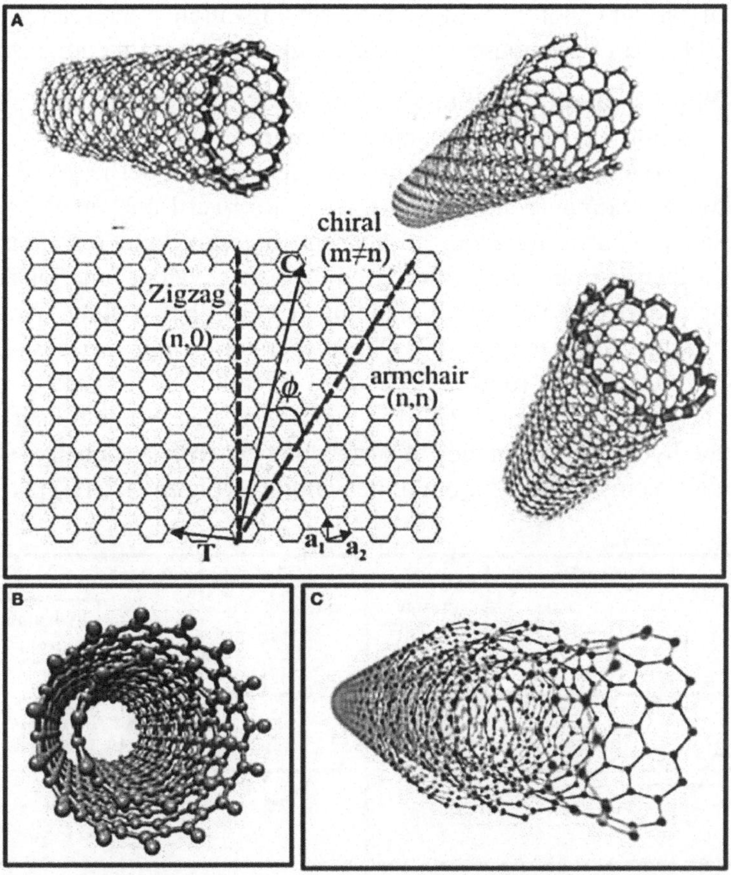

FIGURE 4.1 The structure of a nanotube can be simply specified by the chiral vector, which is given by the orientation of the tube axis with respect to the hexagonal lattice ((A) SWNTs; (B) DWNTs; and (C) MWNTs).

Source: Reprinted with permission from Ref. [11]. https://creativecommons.org/licenses/by/4.0/

4.3 FUNCTIONALIZATION OF CARBON NANOTUBES FOR SENSOR AND ENERGY APPLICATIONS

1. **Exohedral Functionalization:** One of the biggest drawbacks of CNTs for biomedical applications is their inherent difficulty in handling. CNTs do, in fact, tend to form bundles due to strong attractive interactions that are difficult to disrupt. Pristine buckytubes have very hydrophobic surfaces and are insoluble in water and other common solvents as they are produced. The addition of functional groups to the surface

of carbon nanotubes aids in their solubilization and research [20–25]. CNTs can have non-covalent or covalent surface functionalization.

2. **Non-Covalent Functionalization of CNTs:** Non-covalent functionalization has several advantages, the most important of which is the preservation of CNT structural and electrical properties. Additionally, these treatments are typically straightforward and quick, requiring steps such as ultrasonication, centrifugation, and filtration. When surfactant molecules (often used for non-covalent functionalization) are bonded to the carbon nanotube's wall, they assume a variety of configurations (Figure 4.2). The dispersion of nanotubes is stabilized by electrostatic repulsion between micelles in the case of charged surfactants (Figure 4.2(A)); in the case of charge-neutral surfactants, the hydrophilic moieties of surfactants generate a shell-like structure that seems gathered around the nanotube (Figure 4.2(B)) [11].

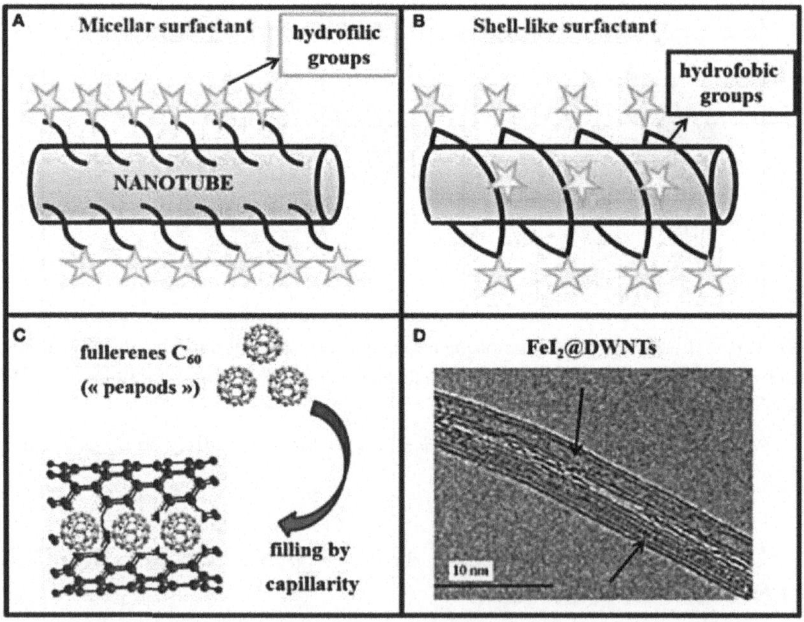

FIGURE 4.2 When surfactant molecules (often used for non-covalent functionalization) are bonded to the carbon nanotube's wall, they assume a variety of configurations ((A) Non-covalent surface with a micellar surfactant; (B) with a shell-like surfactant wrapping around the nanotubes; (C) model of nanotubes filled with "peapods" (C_{60}); and (D) DWNTs filled on all the length with molten FeI_2).

Source: Reprinted with permission from Ref. [11]. https://creativecommons.org/licenses/by/4.0/

4.4 CARBON NANOTUBES FOR SENSOR APPLICATIONS

Carbon nanotubes have been used in a wide range of biosensor applications, utilizing various facets of their beneficial physical and chemical properties. These biosensors are classified based on the sort of transduction mechanism they use. Electrochemical, optical, and field-effect-based sensors are the most prevalent classes [11].

1. **Electrochemical CNT Biosensors:** The Clark electrode was invented in 1962 [26], and electrochemical sensors are the oldest type of biosensor. Because of their relative simplicity, low cost, and portability, electrochemical biosensors are still widely used today. CNTs can be used to improve the performance of electrochemical sensors because of their large surface areas and outstanding electrical characteristics. They have been utilized to boost the sensitivity and decrease the detection limit of electrochemical sensors by improving the immobilization of biological recognition elements, increasing the active surface areas of electrodes, and improving electron transmission [27, 28].
2. **Electrochemical Enzyme Sensor:** Because glucose sensing is so common, electrochemical enzyme sensors are perhaps the most common type of biosensor. CNT-based enzyme sensors have been developed for a variety of applications, including glucose sensing [29, 30], fruit ripening detection [31], and seafood rotting detection [32, 33].
3. **Electrochemical Immunosensors:** Immunosensors are devices that make use of antibody-antigen binding interactions. A sensor surface can be immobilized with either the antibody or the corresponding antigen, and the binding of the corresponding member of the pair, forming a stable complex, can be detected using a variety of methods. Because of the very specific nature of their bonding with their respective antigens, antibodies are commonly used in biosensing applications [34–46], and there are several instances of CNT-immunosensors. The sandwich assay is a typical method in CNT-immunosensors.

   The electrode is functionalized with CNTs in this method, antibodies are immobilized at the electrode—either directly to the CNTs or to another binding element on the electrode surface—and the binding of the analyte proteins is detected by a fluorescent probe [45].
4. **Electrochemical DNA Sensors:** Because of its powerful and very specific hybridization, DNA is an appealing biorecognition

ingredient for sensors. DNA sensors could be used in a variety of ways, such as medicine and healthcare, food safety, and counterterrorism, to name a few [46, 47].

5. **Non-Biomolecule-based Electrochemical Sensors:** While the biomolecule-based sensors discussed in the preceding sections are likely the most popular electrochemical CNT-based sensors, there are numerous others that do not use biomolecules as recognition elements.

Sensors based on molecularly imprinted polymers (MIPs) have gained popularity in recent years. MIPs are synthesized antibodies, which means that polymers with the target molecule bonded either covalently or non-covalently to the polymer structure are synthesized. The target molecule is then removed from the polymer matrix, leaving voids with a certain shape that will allow the target molecule to rebind to it. In the presence of the target molecule in solution, the MIP will selectively bind to it, similar to how antibodies and antigens do [48, 49].

Zhang et al. [50] used MWCNTs to create a sensor for analytes in human perspiration. They created carbon paste electrodes on a flexible polyethylene substrate, and then drop-cast MWCNTs on the electrodes. They also drop-cast MXene-Ti_3C_2Tx on a PET substrate and transferred it to the electrode surface. An ion-selective valinomycin membrane was then applied to the electrode to selectively transfer potassium ions to the electrode surface (Figure 4.3).

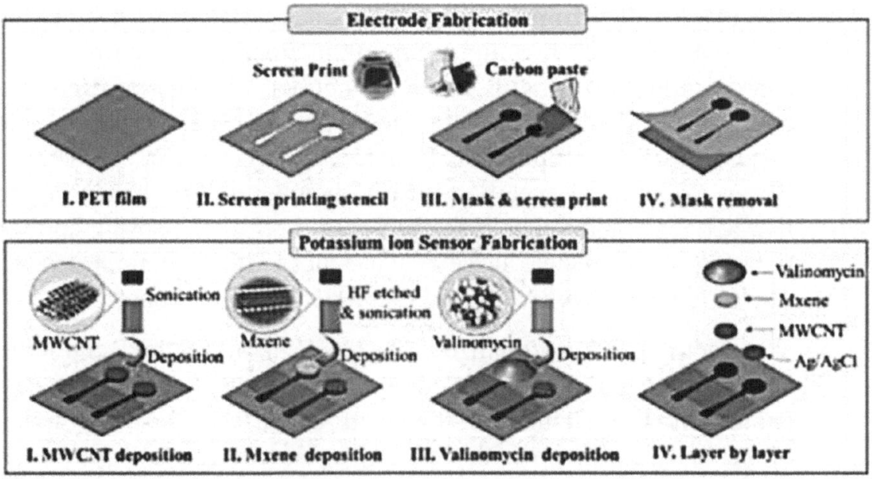

FIGURE 4.3 Fabrication process of MWCNT-MXene potassium ion sensors.

Source: Reprinted with permission from Ref. [50]. Copyright 2021. Elsevier.

CNTs were combined into optical core-shell micro-fibrous textiles by Safaee et al. [51] to make flexible luminous sensors. SWCNTs were non-covalently linked to single-stranded DNA (ssDNA) and distributed in an aqueous solution. The combination of ssDNA-SWCNT is hydrophilic and thus more easily solubilized than native SWCNT due to the ionic DNA binding. Electro spinning was used to mix this dispersion with poly(ethylene oxide) (PEO) and polycaprolactone (PCL) to form core-shell microfibers.

4.5 CONCLUSIONS AND PERSPECTIVES

Carbon nanotubes (CNTs) are a new type of carbonaceous material that have various distinguishing characteristics setting them apart from other carbonaceous materials. Their high chemical stability, distinct and well-defined nanotube tubular structure, surface modification, and strong electrocatalytic activity are some of the qualities that point to their potential. The combination of structural, electrical, and optical qualities in CNTs opens up a world of possibilities for developing new generation nano-tools for sensing applications. CNTs' electronic qualities make them ideal for electrochemical, amperometric, impedometric, or field-effect signal transduction, while their optical properties and ability to pass through biological membranes make them ideal candidates for photoacoustic imaging sensors. The length, diameter, longevity, stability, durability, mechanical qualities, body adjustability, and toxicity must be assessed using a scientific and systematic approach. The high specific area of CNTs is a significant advantage for sensing applications, as it allows for the conjugation or complexation of a large number of receptor moieties, resulting in higher recognition of targets or analytes and, as a result, a lower detection sensitivity threshold. CNTs' capacity to undergo quick electron transfer enables them to overcome most of the shortcomings of traditional electrochemical sensors (such as poor sensitivity and stability, limited repeatability, and long response times for electron transfer reactions). Importantly, CNTs' multifunctionality enables the creation of sensors with a broad range of applications.

KEYWORDS

- **biosensors**
- **carbon nanotube**
- **DNA sensors**
- **electrochemical sensors**
- **electron transfer reactions**
- **immunosensors**
- **sensors**

REFERENCES

1. Saleh, A. A. J., Lee, J. J., & Rahman, M. A., (2009). Electrochemical sensors based on carbon nanotubes. *Sensors, 9*, 2289–2319.
2. Yogeswaran, U., Thiagarajan, S., & Chen, S. M., (2008). Recent updates of DNA incorporated in carbon nanotubes and nanoparticles for electrochemical sensors and biosensors. *Sensors, 8*, 7191–7212.
3. Vashist, S. K., Zheng, D., Al-Rubeaan, L. J. H. T., & Sheu, F. S., (2011). Advances in carbon nanotube based electrochemical sensors for bioanalytical applications. *Biotechnol. Adv., 29*, 169–188.
4. Wang, J., (2005). Carbon-nanotube based electrochemical biosensors: A review. *Electroanalysis, 17*, 7–14.
5. Cha, T. G., Baker, B. A., Sauffer, M. D., Salgado, J., Jaroch, D., Rickus, J. L., Porterfield, D. M., & Choi, J. H., (2011). Optical nanosensors architecture for cell-signaling molecules using DNA aptamer-coated carbon nanotubes. *ACS Nano, 5*, 4236–4244.
6. Chen, Z., Zhang, X., Yang, R., Zhu, Z., Chen, Y., & Tan, W., (2011). Single-walled carbon nanotubes as optical materials for biosensing. *Nanoscale, 3*, 1949–1956.
7. Bandaru, N. M., & Voelcher, N. H., (2012). Glycoconjugate-functionalized carbon nanotubes in biomedicine. *J. Mater. Chem., 22*, 8748–8758.
8. David, C. F., & Kevin, C. H., (2021). Carbon nanotube (CNT)-based biosensors. *Biosensors, 11*, 486. https://doi.org/10.3390/bios11120486.
9. Abdolahad, M., Janmaleki, M., Taghinejad, M., Taghnejad, H., Salehi, F., & Mohajerzadeh, S., (2013). Single-cell resolution diagnosis of cancer cells by carbon nanotube electrical spectroscopy. *Nanoscale, 5*, 3421–3427. doi: 10.1039/c3nr33430a.
10. Abdolahad, M., Taghinejad, M., Taghinejad, H., Janmaleki, M., & Mohajerzadeh, S., (2012). A vertically aligned carbon nanotube-based impedance sensing biosensor for rapid and high sensitive detection of cancer cells. *Lab Chip, 12*, 1183–1190. doi: 10.1039/c2lc21028b.
11. Tîlmaciu, C. M., & Morris, M. C., (2015). Carbon nanotube biosensors. *Front. Chem., 3*, 59. doi: 10.3389/fchem.2015.00059.

12. Ajayan, P. M., & Lijima, S., (1993). Capillarity-induced filling of carbon nanotubes. *Nature, 361*, 333, 334. doi: 10.1038/361333a0.
13. Al-Jamal, K. T., Gherardini, L., Bardi, G., Nunes, A., Guo, C., Bussy, C., et al., (2011). Functional motor recovery from brain ischemic insult by carbon nanotube-mediated siRNA silencing. *Proc. Natl. Acad. Sci. U.S.A., 108*, 10952–10957. doi: 10.1073/pnas.1100930108.
14. Al-Jamal, K. T., Nunes, A., Methven, L., Ali-Boucetta, H., Li, S., Toma, F. M., et al., (2012). Degree of chemical functionalization of carbon nanotubes determines tissue distribution and excretion profile. *Angew. Chem. Int. Ed., 51*, 6389–6393. doi: 10.1002/anie.201201991.
15. Allen, B. L., Kichambare, P. D., Gou, P., Vlasova, I. I., Kapralov, A. A., Konduru, N., et al., (2008). Biodegradation of single-walled carbon nanotubes through enzymatic catalysis. *Nano Lett., 8,* 3899–3903. doi: 10.1021/nl802315h.
16. Antiochia, R., Lavagnini, I., & Magno, F., (2004). Amperometric mediated carbon nanotube paste biosensor for fructose determination. *Anal. Lett., 37*, 1657–1669. doi: 10.1081/AL-120037594.
17. Bhirde, A. A., Patel, V., Gavard, J., Zhang, G., Sousa, A. A., Masedunskas, A., et al., (2009). Targeted killing of cancer cells *in vivo* and *in vitro* with EGF- directed carbon nanotube-based drug delivery. *ACS Nano, 3*, 307–316. doi: 10.1021/nn800551s.
18. Bianco, A., Kostarelos, K., & Prato, M., (2011). Making carbon nanotubes biocompatible and biodegradable. *Chem. Commun., 47*, 10182–10188. doi: 10.1039/c1cc13011k.
19. Biju, V., (2014). Chemical modifications and bioconjugate reactions of nanomaterials for sensing, imaging, drug delivery and therapy. *Chem. Soc. Rev., 43*, 744–764. doi: 10.1039/C3CS60273G.
20. Boghossian, A. A., Zhang, J., Barone, P. W., Reuel, N. F., Kim, J. H., Heller, D. A., et al., (2011). Near-infrared fluorescent sensors based on single-walled carbon nanotubes for life sciences applications. *ChemSusChem., 4*, 848–863. doi: 10.1002/cssc.201100070.
21. Bohunicky, B., & Mousa, S. A., (2011). Biosensors: The new wave in cancer diagnosis. *Nanotechnol. Sci. Appl., 4*, 1–10. doi: 10.2147/NSA.S13465.
22. Bortolamiol, T., Lukanov, P., Galibert, A. M., Soula, B., Lonchambon, P., Datas, L., et al., (2014). Double-walled carbon nanotubes: Quantitative purification assessment, balance between purification and degradation and solution filling as an evidence of opening. *Carbon, 78*, 79–90. doi: 10.1016/j.carbon.2014.06.051.
23. Fernández-Suárez, M., & Ting, A. Y., (2008). Fluorescent probes for super- resolution imaging in living cells. *Nat. Rev. Mol. Cell Biol., 9*, 929–943. doi: 10.1038/nrm2531.
24. Frank, S., Poncharal, P., Wang, Z. L., & Heer, W. A., (1998). Carbon nanotube quantum resistors. *Science, 280*, 1744–1746. doi: 10.1126/science.280.5370.1744.
25. Giepmans, B. N., Adams, S. R., Ellisman, M. H., & Tsien, R. Y., (2006). The fluorescent toolbox for assessing protein location and function. *Science, 312*, 217–224. doi: 10.1126/science.1124618.
26. Clark, L. C., & Lyons, C., (1962). Electrode systems for continuous monitoring in cardiovascular surgery. *Ann. N. Y. Acad. Sci., 102*, 29–45.
27. Hammond, J. L., Formisano, N., Estrela, P., Carrara, S., & Tkac, J., (2016). Electrochemical biosensors and nanobiosensors. *Essays Biochem., 60*, 69–80.
28. Alim, S., Vejayan, J., Yusoff, M. M., & Kafi, A. K. M., (2018). Recent uses of carbon nanotubes & gold nanoparticles in electrochemistry with application in biosensing: A review. *Biosens. Bioelectron., 121*, 125–136.

29. Sanati, A., Jalali, M., Raeissi, K., Karimzadeh, K., Kharaziha, M., & Mahshid, S. S., (2019). A review on recent advancements in electrochemical biosensing using carbonaceous nanomaterials. *Microchim. Acta, 186*, 773. [CrossRef].
30. Zappi, D., Caminiti, R., Ingo, G. M., Sadun, C., Tortolini, C., & Antonelli, M. L., (2017). Biologically friendly room temperature ionic liquids and nanomaterials for the development of innovative enzymatic biosensors. *Talanta, 175*, 566–572.
31. Chen, C., Ran, R., Yang, Z., Lv, R., Shen, W., Kang, F., & Huang, Z. H., (2018). An efficient flexible electrochemical glucose sensor based on carbon nanotubes/carbonized silk fabrics decorated with Pt microspheres. *Sens. Actuators B, 256*, 63–70.
32. Ruhal, A., Rana, J. S., Kumar, S., & Kumar, A., (2012). Immobilization of malate dehydrogenase on carbon nanotubes for development of malate biosensor. *Cell. Mol. Biol., 58*, 15–20.
33. Dervisevic, M., Custiuc, E., Cevik, E., & Senel, M., (2015). Construction of novel xanthine biosensor using polymeric mediator/MWCNT nanocomposite layer for fish freshness detection. *Food Chem., 181*, 277–283.
34. Madhurantakam, S., Babu, K. J., & Rayappan, J. B. B., (2017). Fabrication of mediator-free hybrid nano-interfaced electrochemical biosensor for monitoring cancer cell proliferation. *Biosens. Bioelectron., 87*, 832–841.
35. Lee, Y. M., Kwon, O. Y., Yoon, Y. J., & Ryu, K., (2006). Immobilization of horseradish peroxidase on multi-wall carbon nanotubes and its electrochemical properties. *Biotechnol. Lett., 28*, 39–43.
36. Hrapovic, S., Liu, Y., Male, K. B., & Luong, J. H. T., (2004). Electrochemical biosensing platforms using platinum nanoparticles and carbon nanotubes. *Anal. Chem., 76*, 1083–1088.
37. Magar, H. S., Ghica, M. E., Abbas, M. N., & Brett, C. M. A., (2017). A novel sensitive amperometric choline biosensor based on multiwalled carbon nanotubes and gold nanoparticles. *Talanta, 167*, 462–469.
38. Wilson, T. A., Musameh, M., Kyratzis, I. L., Zhang, J., Bond, A. M., & Hearn, T. W., (2017). Enhanced NADH oxidation using polytyramine/ carbon nanotube modified electrodes for ethanol biosensing. *Electroanalysis, 29*, 1985–1993.
39. Moyo, M., Okonkwo, J. O., & Agyei, N. M., (2014). An amperometric biosensor based on horseradish peroxidase immobilized onto maize tassel-multi-walled carbon nanotubes modified glassy carbon electrode for determination of heavy metal ions in aqueous solution. *Enzyme Microb. Technol., 56*, 28–34.
40. Lin, Y., Lu, F., Tu, Y., & Ren, Z., (2004). Glucose biosensors based on carbon nanotube nanoelectrode ensembles. *Nano Lett., 4*, 191–195.
41. Yu, X., Chattopadhyay, D., Galeska, I., Papadimitrakopoulos, F., & Rusling, J. F., (2003). Peroxidase activity of enzymes bound to the ends of single-wall carbon nanotube forest electrodes. *Electrochem. Commun., 5*, 408–411.
42. Liu, J., Chou, A., Rahmat, W., Paddon-Row, M. N., & Gooding, J. J., (2005). Achieving direct electrical connection to glucose oxidase using aligned single walled carbon nanotube arrays. *Electroanalysis, 17*, 38–46.
43. Tasca, F., Harreither, W., Ludwig, R., Gooding, J. J., & Gorton, L., (2011). Cellobiose dehydrogenase aryl diazonium modified single walled carbon nanotubes: Enhanced direct electron transfer through a positively charged surface. *Anal. Chem., 83*, 3042–3049.

44. Felix, F. S., & Angnes, L., (2018). Electrochemical immunosensors—A powerful tool for analytical applications. *Biosens. Bioelectron., 102*, 470–478.
45. Silva, M. M. S., Dias, A. C. M. S., Silva, B. V. M., Gomes-Filho, S. L. R., Kubota, L. T., Goulart, M. O. F., & Dutra, R. F., (2015). Electrochemical detection of dengue virus NS1 protein with a poly(allylsamine)/carbon nanotube layered immunoelectrode. *J. Chem. Technol. Biotechnol., 90*, 194–2020.
46. Riberi, W. I., Tarditto, L. V., Zon, M. A., Arevalo, F. J., & Fernandez, H., (2018). Development of an electrochemical immunosensor to determine zearalenone in maize using carbon screen printed electrodes modified with multi-walled carbon nanotubes/polyethyleneimine dispersions. *Sens. Actuators B, 254*, 1271–1277.
47. Ferrier, D. C., Shaver, M. P., & Hands, P. J. W., (2015). Micro-and nanostructure-based oligonucleotide sensors. *Biosens. Bioelectron., 68*, 798–810.
48. Beyazit, Bui, Haupt, K., & Gonzato, C., (2016). Molecularly imprinted polymer nanomaterials and nanocomposites by controlled/living radical polymerization. *Prog. Polym. Sci., 62*, 1–21.
49. BelBruno, J. J., (2019). Molecularly imprinted polymers. *Chem. Rev., 119*, 94–119.
50. Zhang, S., Zahed, M. A., Sharifuzzaman, M., Yoon, S., Hui, X., Barman, S. C., Sharma, S., Yoon, S., Yoon, H. S., Parl, C., et al., (2021). A wearable battery-free wireless and skin-interfaced microfluidics integrated electrochemical sensing patch for on-site biomarkers monitoring in human perspiration. *Biosens. Bioelectron., 175,* 112844.
51. Safee, M. M., Gravely, M., & Roxbury, D., (2021). A wearable optical microfibrous biomaterial with encapsulated nanosensors enables wireless monitoring of oxidative stress. *Adv. Funct. Mater., 31,* 2006254.
52. Ferrari, M., (2005). Cancer nanotechnology: Opportunities and challenges. *Nat. Rev. Cancer, 5*, 161–171. doi: 10.1038/nrc1566.
53. Gao, H., Kong, Y., Cui, D., & Ozkan, C. S., (2003). Spontaneous insertion of DNA oligonucleotides into carbon nanotubes. *Nano Lett., 3*, 471–473. doi: 10.1021/nl025967a.

CHAPTER 5

Analysis of Phase Transitions Based on Free Energy Potential and Interface Thickness

S. L. DIVYA,[1] S. L. NAVYA,[2] and SUNSU KURIAN THOTTIL[3]

[1]*Department of Computer Applications, SCMS School of Technology and Management, Muttom, Kochi, Ernakulam, Kerala, India*

[2]*Department of Physics, Sacred Heart College (Autonomous), Thevara, Kochi, Ernakulam, Kerala, India*

[3]*Department of Physics, St. Teresa's College (Autonomous), Ernakulam, Kerala, India*

ABSTRACT

Phase transitions are observed extensively in physical phenomena such as the condensation of gases, the solidification of liquids, etc. These transitions occur only if certain conditions are met. Therefore, adequate mathematical modeling is required to understand these phenomena. The main objective of this chapter is to study the variations in interface thickness during phase transitions. The widely used mathematical model for this purpose is the phase field model. This model is adopted here to find the correlation between the phase field variable and the free energy potential in the case of solid-liquid phase transition.

5.1 INTRODUCTION

Matter can exist in four states: solid, liquid, gas, and plasma. There are also many intermediate states. A state of matter can be changed to another by varying certain external conditions like temperature, pressure, etc. This

Sustainable Environmental Engineering: Water Security, Energy Conservation, and Green Process.
Sonia Khanna, Ann Rose Abraham, Kian Hariri Asli, & A. K. Haghi (Eds.)

process of transition from one state to another is called phase transition. In this chapter, we consider solid-liquid phase transition [1, 10, 11, 15]. The solid-liquid interface can be described as an evolving phase transition controlled by the boundary conditions. These boundary conditions are consistent with the transformation. When phase transition [2, 12, 13] occurs, the interface thickness approaches zero. So sharp interface conditions are expected in the mathematical modeling of solid-liquid phase transitions. The model used in this study is the phase field model.

5.2 PHASE FIELD MODEL

The phase field model [3, 9, 21–25] is an important mathematical model used to visualize the phase transitions. This method can treat many physical phenomena simultaneously consistent with irreversible thermodynamics. In this model, the phase transition is represented continuously by a parameter known as phase field variable 'p.' The variable p has a value –1 for solid phase and +1 for liquid phase. The intermediate values of p can be represented as 'ξ.' The width of the transition region between solid and liquid phases is controlled by ξ. So we can say that ξ represents the interface thickness. When $\xi \rightarrow 0$, we can consider sharp interface models [28].

Let us consider the heat equation corresponding to the phase transition:

$$\frac{\partial T}{\partial t} = c\frac{\partial^2 T}{\partial \xi^2} \tag{1}$$

where; 'T' represents the temperature at interface thickness ξ, time 't' and constant 'c.'

The solution to this heat equation corresponding to the initial condition:

$$T(\xi,0) = \{a\ for\ \xi > b\ 0\ elsewhere,$$

can be written as:

$$T(\xi,\mathrm{t}) = \frac{a}{2}\left(erf\left(\frac{\xi - b}{4kt}\right)\right) \tag{2}$$

where; 'k' is the heat coefficient of the material.

The phase field mathematical model [4, 16] assumes that phase change takes place over a finite temperature range $[T_s, T_l]$, which corresponds to the transition region of the solid-liquid mixture. At temperature T_s, the state of the material is purely solid, and at temperature T_l, the state is purely liquid. The transition region can be made narrow or wide according to our requirements. Additionally, we can say that the interface thickness is related to the temperature.

Figure 5.1 shows the variation of interface thickness ξ, with respect to the temperature T and time t. It is clear that as time t increases, the interface thickness increases.

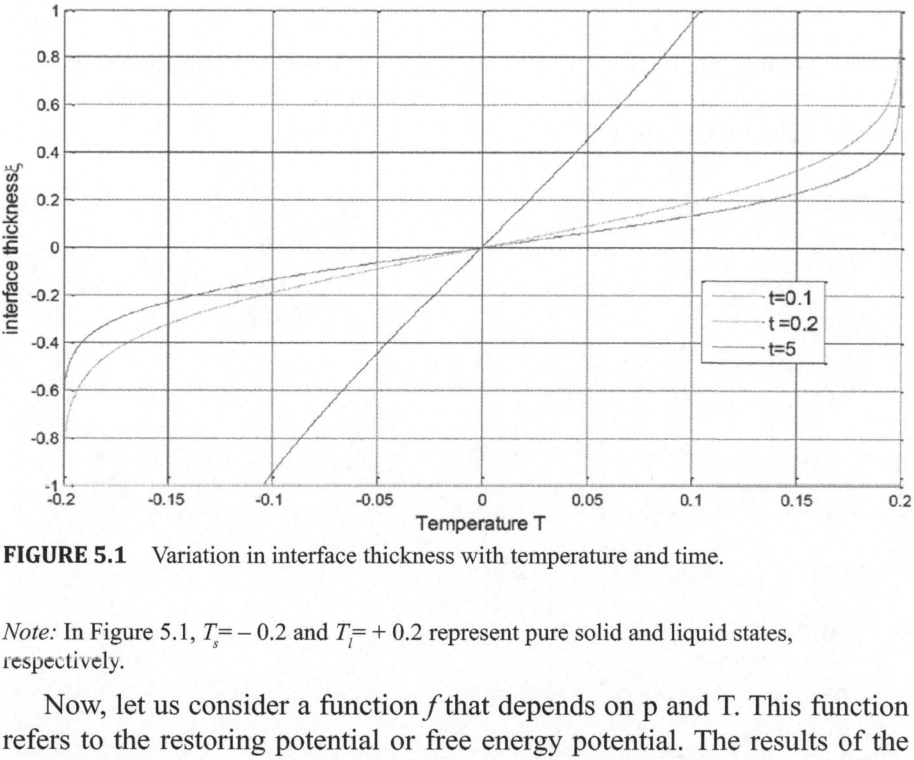

FIGURE 5.1 Variation in interface thickness with temperature and time.

Note: In Figure 5.1, $T_s = -0.2$ and $T_l = +0.2$ represent pure solid and liquid states, respectively.

Now, let us consider a function *f* that depends on p and T. This function refers to the restoring potential or free energy potential. The results of the model discussed here depend largely on the thermodynamic consistency of $f(p,T)$ [1, 14]. For all forms of $f(p,T)$ where T and p coupling is linear, the sharp interface limit is attained. The entropy/energy/temperature scales for a given specific form of $f(p,T)$ the following relationships must be satisfied.

$$\Delta S = (S_{liquid} - S_{solid}) = \frac{\partial f}{\partial T}\bigg|_{solid}^{liquid} \tag{3}$$

$$\Delta S\bigg|_{T=0} = \frac{L_f}{T_m} \tag{4}$$

where; 'S' is entropy; 'ΔS' is change in entropy; 'T_m' is mean temperature; and 'L_f' is latent heat [5, 17–20].

According to the Caginalp Potential (CP) model [6], the dependence of f on T can be written as:

$$f(p,T) = \frac{1}{8a}(p^2 - 1)^2 - \frac{\Delta S}{2} pT \tag{5}$$

The parameter 'a' is chosen in such a way that the partial derivative $\frac{\partial f}{\partial p}$ exhibits three roots near 0 and ±1. As T depart from zero, the minima of $f(p,T)$ at $p = \pm 1$ changes (Figure 5.2).

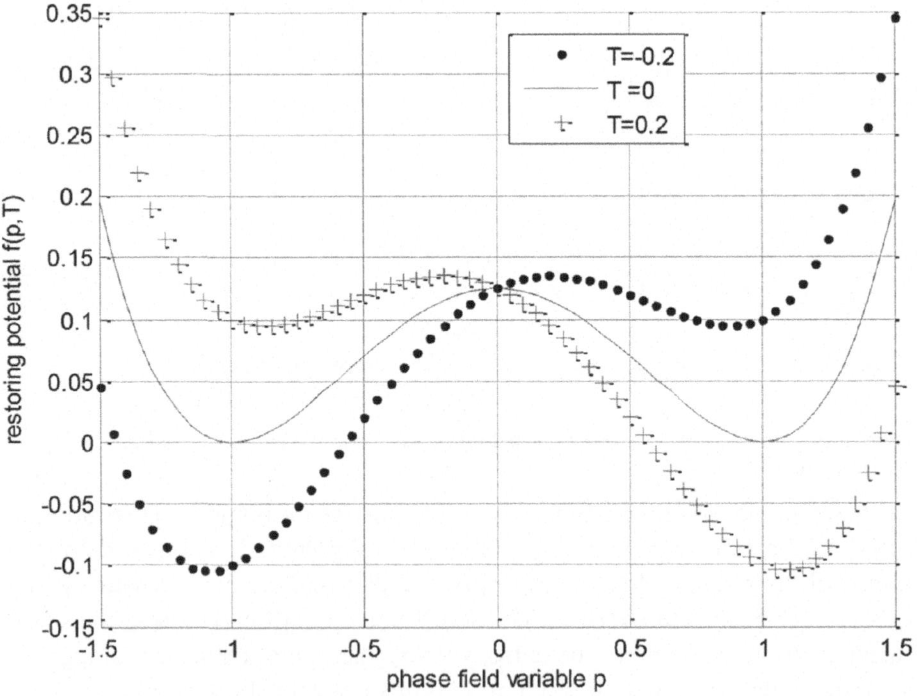

FIGURE 5.2 Plot showing the relation between phase field variable and restoring potential at temperature T.

When p = 0, the restoring potential $f(p,T)$ is a constant. Also when $T = 0$, the variation of restoring potential with phase field variable is symmetric and the phases are equally stable.

Figures 5.3 and 5.4 show the variation of the restoring potential with time and phase field variable for a constant heat transfer coefficient. It is clear that the variation of $f(p,T)$ is symmetric in both cases. But, the minimum value of restoring potential for both solid and liquid varies with time and the parameter 'a.'

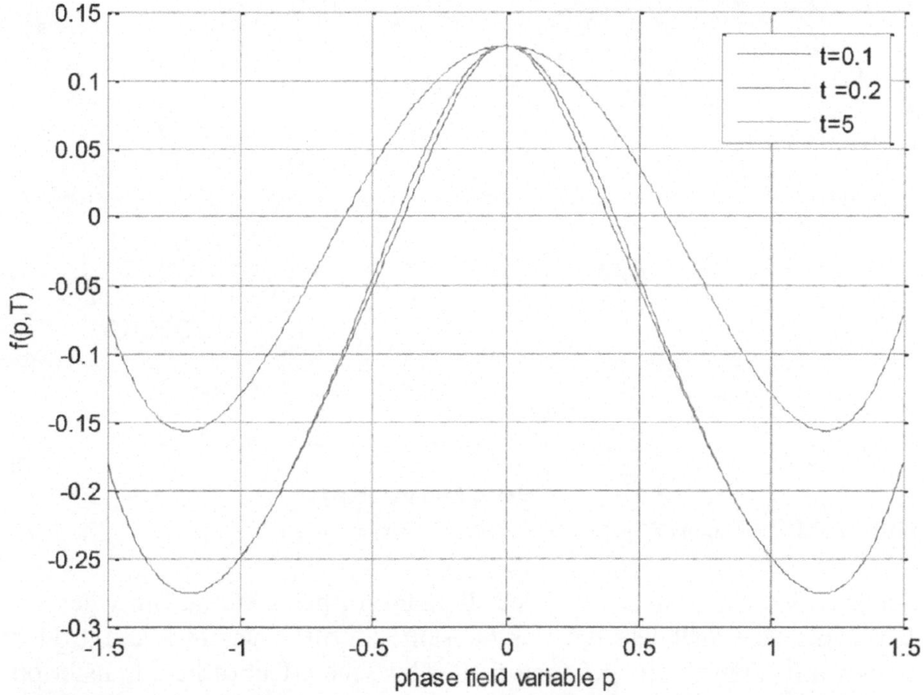

FIGURE 5.3 Variation of f(p,T) with p for a = 1 and k = 0.2.

5.3 HEAT TRANSFER COEFFICIENT

Heat is a form of energy produced by the variation in temperature. A phase change occurs when we add or remove heat from a system. Also, different materials have different heat resistance. This can be analyzed from the heat transfer coefficient. Heat transfer coefficient [7, 26, 27, 29, 30] describes how well heat is transferred from a solid to a liquid or gas (vice versa).

5.4 VARIATION OF INTERFACE THICKNESS BASED ON HEAT TRANSFER COEFFICIENT

We study the variation of interface thickness with heat transfer coefficient. From Figures 5.5, 5.7, and 5.9, it is clear that when t = 0.1, the interface

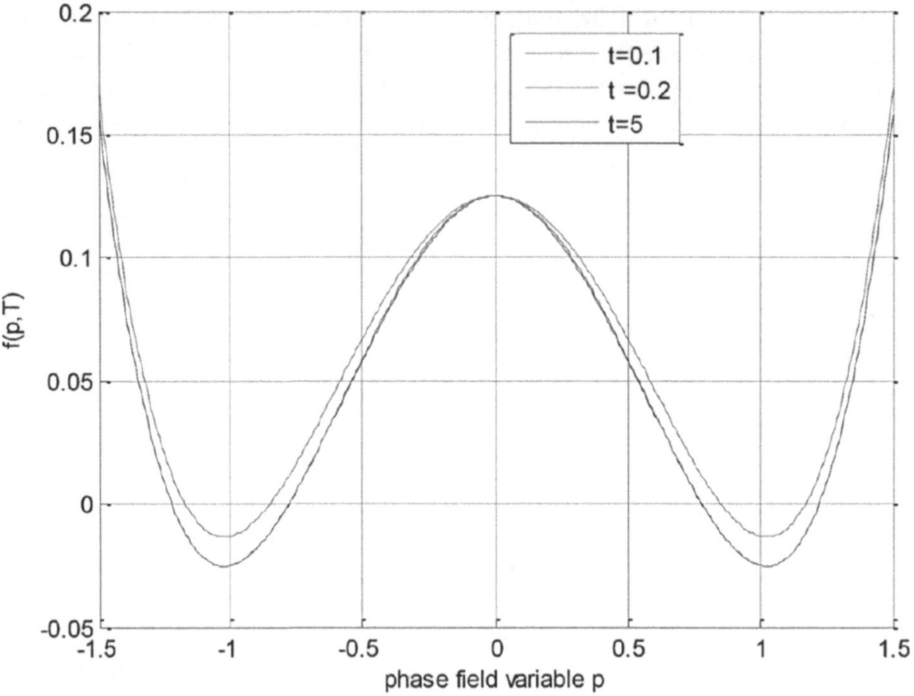

FIGURE 5.4 Variation of f(p,T) with p for a = 0.1 and k = 0.2.

thickness is maximum at k = 1 for all values of parameter a. But when t = 1, the interface thickness remains the same for all values of k and a. This is shown in Figures 5.6, 5.8, and 5.10. Thus, the effect of heat transfer on interface thickness decreases with time t.

In the next section, we discuss the correlation between the phase field variable and free energy potential.

5.5 CORRELATION BETWEEN PHASE FIELD VARIABLE AND FREE ENERGY POTENTIAL

We use correlation analysis to check the covariation of phase field variable p and free energy potential $f(p,T)$. There are three possibilities:

i. p and f may increase or decrease together;
ii. One of them increases while the other decreases; and
iii. The variation in one variable has no relation with that of the other.

In case (i), we can say that the variables p and f are positively correlated or directly correlated. In case (ii), they are negatively correlated or inversely

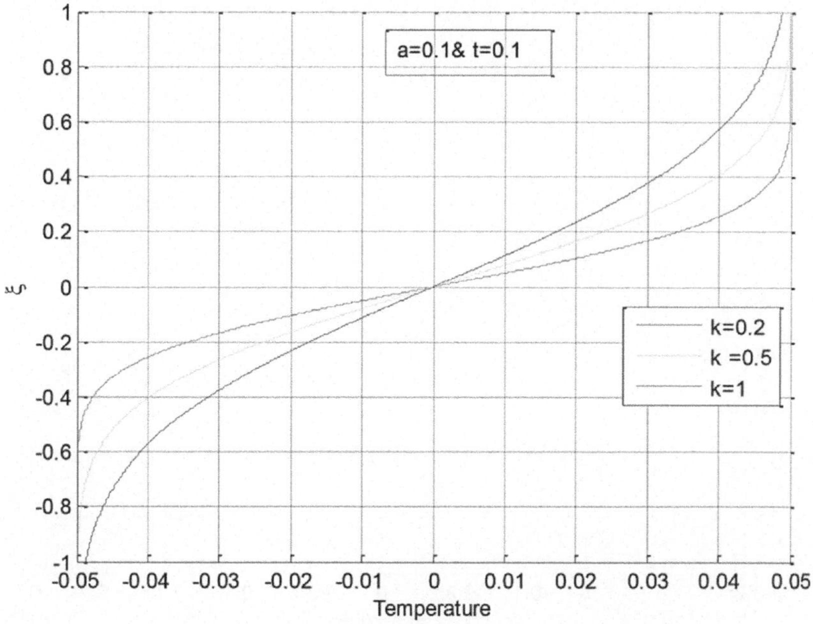

FIGURE 5.5 Plot showing the variation of interface thickness with temperature when a = 0.1 and t = 0.1.

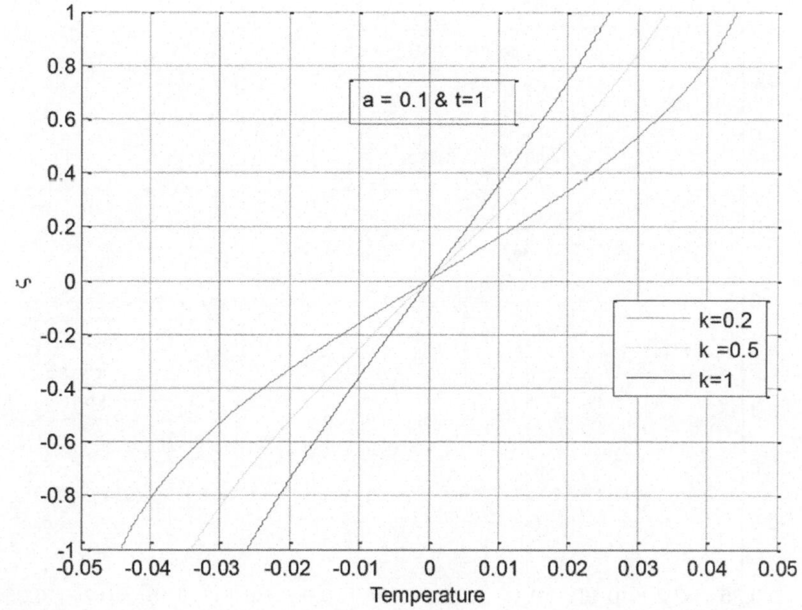

FIGURE 5.6 Plot showing the variation of interface thickness with temperature when a = 0.1 and t = 1.

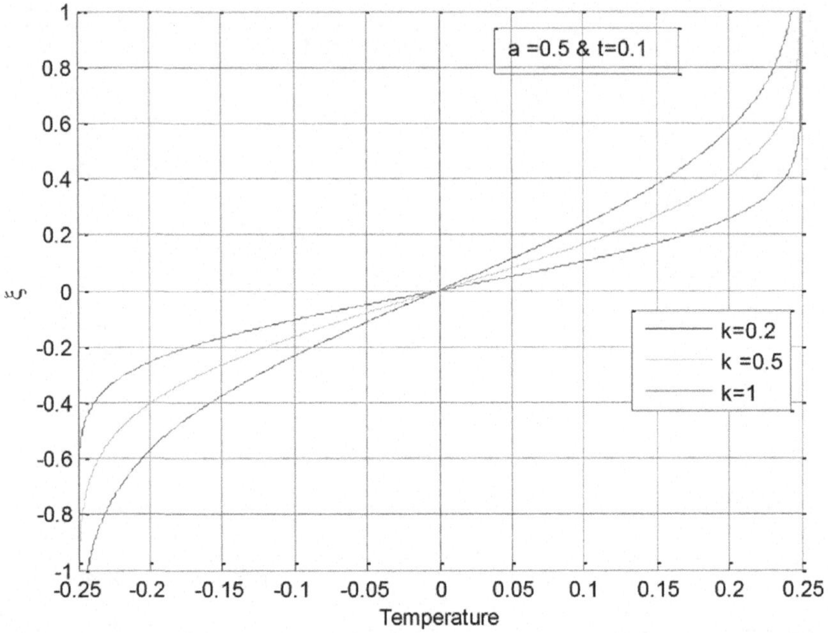

FIGURE 5.7 Plot showing the variation of interface thickness with temperature when a = 0.5 and t = 0.1.

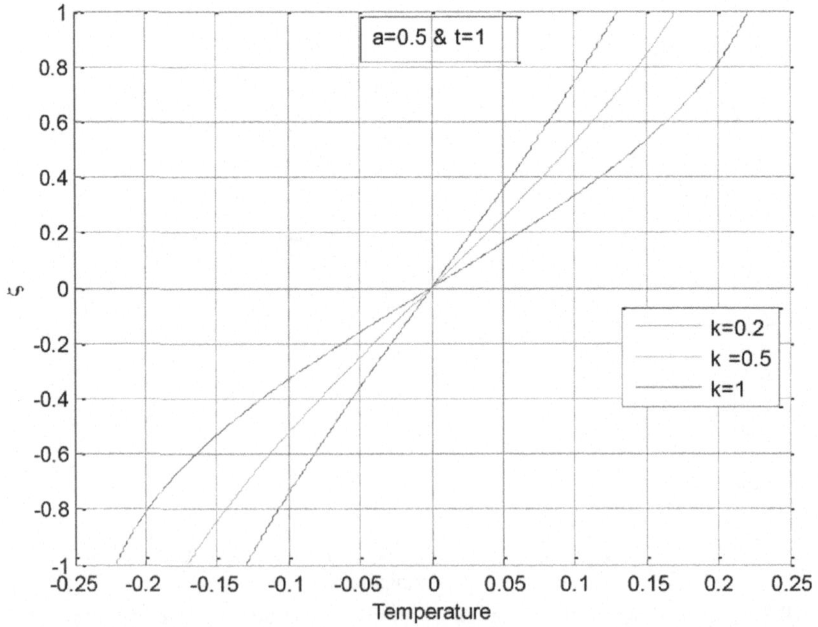

FIGURE 5.8 The variation of interface thickness with temperature when a = 0.5 and t = 1.

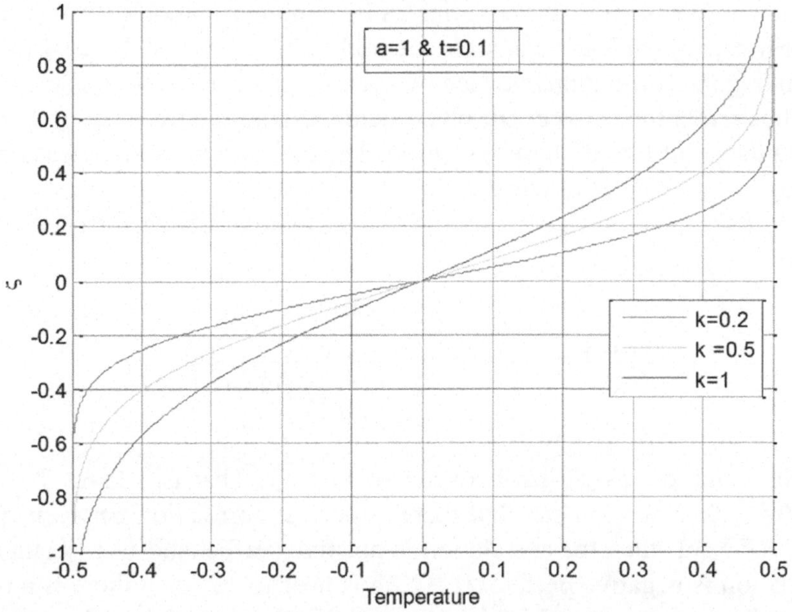

FIGURE 5.9 The variation of interface thickness with temperature when a = 1 and t = 0.1.

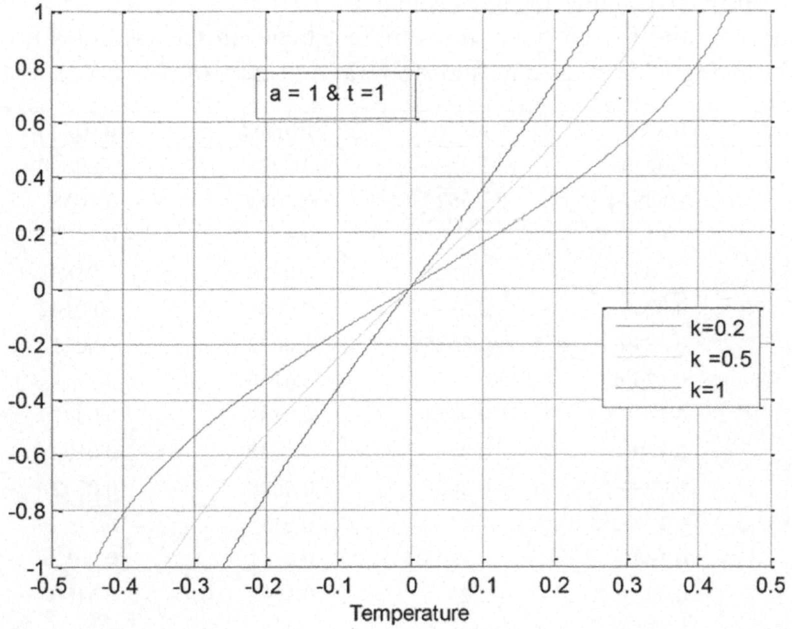

FIGURE 5.10 Plot showing the variation of interface thickness with temperature when a = 1 and t = 1.

correlated. In case (iii), these variables have no correlation or zero correlation between them.Karl Pearson's method is the most widely used method for finding the correlation between two variables. Here we use this method to find the correlation between the phase field variable and free energy potential during the solid-liquid phase transition. The correlation can be measured by finding the coefficient of correlation.

The Karl Pearson's coefficient [8] of correlation between two variables X and Y is given by:

$$r(X,Y)=\frac{\Sigma x_i y_i-\frac{\Sigma x_i \Sigma y_i}{n}}{\sqrt{\left[\Sigma x_i^2-\frac{(\Sigma x_i)^2}{n}\right]\left[\Sigma y_i^2\frac{(\Sigma y_i)^2}{n}\right]}}$$

The value of $r(X,Y)$ lies between –1 and +1. That is, $-1\leq r(X,Y)\leq 1$. If $r(X,Y)=\pm 1$,then we can say that there is perfect correlation between X and Y. If $r(X,Y)=1$, then the correlation is positive perfect and if it –1, then the correlation is negative perfect. If $r(X,Y)$ is near to +1, then there is a direct correlation between the variables X and Y. If $r(X,Y)$ is near to –1, there is an inverse correlation between X and Y and if $r(X,Y)=0$, then there is no linear correlation between the variables X and Y.

Here we find the correlation coefficient between the variables p and f. The following table is used to find $r(p,f)$ at $a=1, \Delta S=1 \& T=-0.2$.

p	**f(p,T)**	**p2**	**f(p,T)2**	**p.f(p,T)**
–1.5	0.0453	2.25	0.0021	–0.0680
–1.25	–0.0854	1.5625	0.0073	0.1068
–1	–0.1	1	0.01	0.10
–0.75	–0.0511	0.5625	0.0026	0.0383
–0.5	0.0203	0.25	0.0004	–0.0102
–0.25	0.0849	0.0625	0.0072	–0.0212
0	0.1250	0	0.0156	0
0.25	0.1349	0.0625	0.0182	0.0337
0.5	0.1203	0.25	0.0145	0.0602
0.75	0.0989	0.5625	0.0098	0.0742
1	0.1	1	0.01	0.1
1.25	0.1646	1.5625	0.0271	0.2057
1.5	0.3453	2.25	0.1192	0.5180
$\sum p = 0$	$\sum f(p,T) = 1.0029$	$\sum p^2 = 11.3750$	$\sum f(p,T)^2 = 0.2440$	$\sum p.f(p,T) = 1.1375$

$$r(p,f)=\frac{\Sigma pf-\dfrac{\Sigma p\Sigma f}{n}}{\sqrt{\left[\Sigma p^2-\dfrac{(\Sigma p)^2}{n}\right]\left[\Sigma f^2\dfrac{(\Sigma f)^2}{n}\right]}}$$

$$=0.8263$$

In this case, there is a direct correlation between p and f. Similarly, we find that the correlation between p and f when $a=1, \Delta S=1 \& T=0$ is $r(p,f)=0$. Hence there is no linear correlation between p and f in this case. Also for $a=1, \Delta S=1 \& T=-0.2.$, $r(p,f)=-0.8263$, which shows an inverse correlation between p and f.

5.6 CONCLUSION

Phase field model is an excellent tool to study phase transitions. In this chapter, we discussed the variation of restoring potential with phase field variable. This restoring potential depends on the heat transfer coefficient. Also, the intermediate values of phase field variable can be considered as the interface thickness. Thus, we studied the variation in interface thickness with heat transfer coefficient and temperature. It is clear that the interface thickness increases with heat transfer coefficient and temperature. We also analyzed the correlation between phase field variable and restoring potential. This model can also be used to study the growth of the precipitate which is isolated by an interface.

KEYWORDS

- **correlation**
- **entropy**
- **heat transfer coefficient**
- **inverse correlation**
- **phase field variable**
- **restoring potential**

REFERENCES

1. Surana, K. J., (2015). Mathematical models and numerical solutions of liquid-solid and solid-liquid phase change. *Journal of Thermal Engineering*, 61–98.
2. Avrami, M., (1939). Kinetics of phase change 1. *J. Chem. Phys*, 1103–1112.
3. Bhadeshia, R. S., (2009). Phase-field model study of the effect of interface anisotropy on the crystal morphological evolution of cubic metals. *Acta Mater, 57*, 2210–2216.
4. Tan, Z., & Lim, K. M., (2007). An adaptive mesh redistribution method for the incompressible mixture flows using phase-field model. *J. Comput. Phys. 225*, 1137–1158.
5. Verma, P., Varun & Singal, S. K., (2008). Review of mathematical modeling on latent heat thermal energy storage systems using phase-change material. *Renewable and Sustainable Energy Reviews, 12*, 999–1031.
6. Voller, V., & Fabbri M., (1997). The phase-field method in sharp interface limit: A comparison between model potentials. *Journal of Computational Physics*, 256–265.
7. Sukhatma, S. P., (2005). *A Text Book on Heat Transfer.* Universities Press.
8. Sharma, A. K., (2005). *Text Book of Correlation and Regression.* Discovery Publishing House.
9. Alexiades, V., & Solomon A. D., (1993). *Mathematical Modeling of Melting and Freezing Processes.* USA: Hemisphere Publishing Corporation.
10. Avrami, M., (1940). Kinetics of phase change 2. *J. Chem. Phys*, 212–224.
11. Avrami, M., (1941). Kinetics of phase change 3. *J. Chem. Phys*, 177–184.
12. Bilir, L., & Ilken, Z., (2005). Total solidification time of a liquid phase change material enclosed in cylindrical/spherical containers. *Applied Thermal Engineering, 25*, 1488–1502.
13. Bonacina, C., (1973). Numerical solution of phase-change problems. *Int. J. Heat Mass Transfer*, 1825–1832.
14. Swaminathan, C., & Voller, V., (1993). On the enthalpy method. *Int J Num Meth, Heat Fluid Flow, 3*, 233–244.
15. Cao, Y., & Faghri, A., (1990). A numerical analysis of phase change problem including natural convection. *J. Heat Transfer, 112*, 812–815.
16. Costa, M., & A. O., (1991). Numerical simulation of solid–liquid phase change phenomena. *Comput. Meth. Appl. Mech. Eng. 91*, 1123–1134.
17. Costa, M., & D. B., (1998). Numerical simulation of a latent heat thermal energy storage system with enhanced heat conduction. *Energy Conversion and Management, 39*(3, 4), 319–330.
18. Hunter, L. W., & Kuttler, J. R., (1989). The enthalpy method for heat conduction problems with moving boundaries. *J. Heat Transfer, Trans. ASME, 111*, 239–242.
19. Arnold, V. R., (1989). *Mathematical Methods of Classical Mechanics.* New York: Springer.
20. Jaeger, H. S., (1959). *Conduction of Heat in Solids.* Clarendon Press, Oxford.
21. Jaluria, R. V., (1993). A comparison of different solution methodologies for melting and solidification problems in enclosures. *Numer. Heat Transfer, Part B: Fundam. 24*, 77–105.
22. Mondal, S., (2008). Phase change materials for smart textiles – An overview. *Applied Thermal Engineering, 28*, 1536–1550.

23. Morgan, K., (1981). A numerical analysis of freezing and melting with convection. *Comp. Meth. App. Eng., 28*, 275–284.
24. Nedjar, B., (2002). An enthalpy-based finite element method for nonlinear heat problems involving phase change. *Comput. Struct., 80*, 9–21.
25. Qin, R. S., & Bhadeshia, H. K., (2010). Review: Phase field method. *Materials Science and Technology*, 803–811.
26. Suhas, V. P., (1980). *Numerical Heat Transfer and Fluid Flow.* Washington, DC: Hemisphere Publishing Corporation.
27. Sharma, et al., (2009). Review on thermal energy storage with phase change materials and applications. *Renewable and Sustainable Energy Review, 13*(2), 318–345.
28. Socolovsky, & Caginalp, G., (1991). Computation of sharp phase boundaries by spreading: The planar and spherically symmetric cases. *Journal of Computational Physics, 95,* 85–100.
29. Voller, V. R., (1987). An enthalpy method for convection/diffusion phase change. *Int. J. Numer. Meth. Eng., 24*, 271–284.
30. Voller, V. R., (1990). Fast implicit finite-difference method for the analysis of phase change problems. *Numer. Heat. Transfer Part B, 17,* 155–169.

CHAPTER 6

Nanoparticles for Sustainable Water Purification and Dye Removal

K. B. AKHILA and RONY RAJAN PAUL

Department of Chemistry, CMS College, Kottayam, Kerala, India

ABSTRACT

Water is a vital resource for the existence of mankind. It is the spirit of life in our universe. The efficient and targeted eradication of organic dye from water systems is a critical global concern for the treatment of both drinking water and wastewater [1]. Dyes pose major environmental, aesthetic, and health issues since they are persistent, stubborn colorants, and even possibly carcinogenic and poisonous. Industrial effluents containing dyes are a growing source of concern and must be properly treated before being released into the environment to avoid these possible risks [2]. High surface area, photosensitivity, catalytic, and antibacterial activity, electrochemical, optical, and magnetic capabilities, as well as adjustable properties, are some of the distinctive characteristics of nanomaterials [3] that account for their application in water pollution prevention, detection, water quality monitoring, and remediation of pollutants. In this chapter, we are evaluating the application of various nanoparticles in water purification and dye removal.

6.1 INTRODUCTION

One of the biggest problems that the world faces with the evolution of different industries is water pollution [4]. Organic dyes and water are used in the dyeing process in various industries, including textile, leather, paper, cosmetics, plastics, food, and others. Disposal of dyes, which are actually brightly colored organic chemicals that aren't very biodegradable, into water

Sustainable Environmental Engineering: Water Security, Energy Conservation, and Green Process.
Sonia Khanna, Ann Rose Abraham, Kian Hariri Asli, & A. K. Haghi (Eds.)

resources is a significant contributor to water pollution. The textile sector produces wastewater that is brightly colored and generally has a concentration between 10 and 200 mg/L [2]. Dyes offer major environmental, aesthetic, and health issues, since they are persistent, stubborn colorants and even possibly carcinogenic and poisonous. Industrial dye-laden effluents are therefore growing problems that require appropriate treatment before being released into the environment to avoid these possible risks. Persistent organic compounds present in cationic and anionic dyes are extremely challenging to break down into nontoxic compounds. In general, dye molecules are resistant to heat, light, chemical oxidizers, aerobic digestion, biological, and photochemical degradation. However, these dyes have a negative impact on fish, people, plants, and other living beings. Based on these adverse effects, dye molecules are grouped under organic pollutants. Due to the dyeing industry's high effluent discharge rates into water bodies, water pollution caused by its operations is a serious problem. The makeup of dye effluents varies, and it may be quite complicated and poisonous in nature [6].

Numerous conventional methods have been used to remove toxic pollutants from water and wastewater due to an increase in pollutants and a rapid rate of industrialization, including coagulation, membrane processes, adsorption, dialysis, advanced oxidation processes, photocatalytic degradation, and biological methods. The cost-effectiveness and energy requirements of these systems make it difficult to fulfill the high standards for water quality. Therefore, new technologies must be developed in order to provide clean and accessible water to many regions of the world and to overcome the problems caused by water pollution [3].

Adsorption, most likely due to its simplicity and effectiveness, is one of the significant methods for dye removal. Different kinds of adsorbents, including organic and inorganic materials, have been used for the removal of dyes from colored water and wastewater. One of these materials that has been the subject of extensive research and usage as an adsorbent is activated carbon. However, due to its high manufacturing and processing costs, several scholars have focused their efforts on creating innovative alternative adsorbents with high adsorptive capacity and low cost, as well as optimizing adsorption. In this context, nanotechnology has recently gained a lot of interest [5].

Among the top five revolutionary technologies for water purification is nanotechnology. The distinctive qualities of nanomaterials, such as large surface area, antimicrobial activity, catalytic activity, photosensitivity, optical and magnetic properties, tunable pore size, and surface chemistry, provide a wide range of applications in water pollution prevention, detection, and water quality monitoring.

6.2 WATER POLLUTION AND DYES

Many organic substances, including pesticides, organochlorines, polychlorinated biphenyls, polycyclic aromatic hydrocarbons, and organic dyes, are significant water pollutants. There is widespread usage of very poisonous and vibrant organic dyes across a number of sectors, including textile, paper, leather, cosmetics, and dyestuff businesses [9]. Concern should be expressed over how polluted effluents affect aquatic life, vegetation, people's health, and many economic sectors. Unwanted ecological change is accompanied by the discharge of untreated industrial and home effluents into the environment [1]. The severity of the issues caused by the dumping of industrial effluents into the environment, especially how they affect people's health, resulted in increased attention on research and development in the area of industrial water treatment [6]. As dyes are persistent organic compounds that are resistant to aerobic digestion and not susceptible to photolytic decomposition, dye-containing wastewater is highly challenging to treat. Due to their synthetic origin and intricate chemical composition, they are persistent in nature. Dye compounds' mutagenic and carcinogenic properties are becoming a serious concern for the environment. It was discovered that basic and diazo dyes are more toxic. In aquatic environments, the color of dyes is highly noticeable and affects the clarity, esthetics, and gas solubility of the water [9].

These pollutants have recently gained significant scientific attention, and removing them from wastewater directly impacts people's quality of life. There is an increasing demand for the removal of toxins, carcinogens, and mutagenic substances found in wastewater effluents. The removal of dye effluents from the environment has been achieved through chemical treatment techniques such as adsorption, filtration, coagulation-flocculation, reverse osmosis in an aqueous environment, ion-exchange adsorption, osmosis, precipitation, chlorine, and hydrogen peroxide treatment [9].

6.3 NANOPARTICLES FOR WATER PURIFICATION

6.3.1 MAGNETIC NANOPARTICLES

Due to their high pollutant removal capacity, quick kinetics, and reactivity, as well as their high separation efficiency and reusability, magnetic nanoparticles have attracted considerable attention [8]. The work by Y-R. Zhang, S-Q. Wang, S-L. Shen, and B-X. Zhao describes the preparation and adsorption behavior of Fe_3O_4 magnetic nanoparticles (MNPs) modified with 3-glycidoxypropyltrimethoxysilane (GPTMS) and glycine (Gly).

Based on the electrostatic adsorption theory, they developed and produced double-function magnetic nanoparticles utilizing glycine and low toxicity organosilane as the starting components. Both cationic and anionic dyes are effectively removed from water solutions by MNPs. They created the magnetic adsorbent (Fe_3O_4@GPTMS@Gly), and TEM, XRD, IR, and TGA were used to characterize the adsorbent. The MNPs can effectively remove both cationic dyes (MEB and AZ) and anionic dyes (MB, OR, and AR) in extreme circumstances from aqueous solutions [8].

The removal of Rhodamine B, a common hydrophilic organic dye used in industry, was accomplished in the work of Amiralian et al. employing magnetic nanocellulose membranes with varying contents of magnetic nanoparticles and enhanced oxidation procedures using active groups of sulfate and oxygen [10]. It was demonstrated that the Fe_3O_4 nanoparticles grafted onto the CNF are in which magnetic nanoparticles are initially formed and proliferated before appearing on the surface of nanofibers. High magnetic properties are a result of the structure of iron oxide nanoparticles, which is tightly packed on the surfaces of nanofibers. It has been shown that the magnetic nanocellulose membrane is very efficient at catalyzing the removal of RhB by PMS. Also, this research offers fresh perspectives towards the construction of flexible magnetic nanocellulose membranes with excellent magnetic characteristics for possible wastewater treatment applications [10] (Figure 6.1).

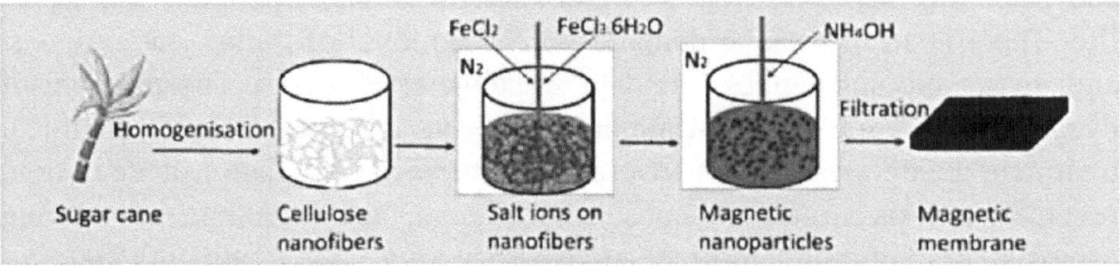

FIGURE 6.1 Preparation of magnetic nanocellulose membrane.

Source: Reprinted with permission from Ref. [10]. Crown Copyright © 2020 Published by Elsevier B.V.

6.3.2 METAL OXIDE NANOPARTICLES

One of the most promising methods for addressing environmental problems with wastewater contaminated by dyes is heterogeneous photocatalysis employing semiconductive metal oxides such as TiO_2 and ZnO [11]. Due to their intriguing properties, including low cost, non-toxicity, high stability, strong photocatalytic activity, and capacity to facilitate ambient temperature oxidation, titanium dioxide (TiO_2) is the most researched metal oxide. Due to its high excitation binding energy (60 meV) at ambient temperature, great photosensitivity, low cost, and comparable band gap to TiO_2, zinc oxide (ZnO) is the second most extensively researched catalyst for water depollution (3.2 eV). Colloidal TiO_2 and ZnO nanoparticles have the attractive advantages of low-cost, large-scale solution processing and controllable optoelectronic characteristics via controlled synthesis and surface chemical engineering. Nonetheless, they are required to be trapped on a porous support because of recycling problems. Heterogeneous catalysis has made considerable use of mesoporous silica as a porous support. Organic dyes can be removed from wastewater using adsorption (dark adsorption) using a core/shell structured ZnO/mesoporous silica with ZnO as a shell photocatalyst, with the potential for regeneration of the dye-adsorbed photocatalysts via degradation of organic dyes (photocatalytic reaction under UV irradiation).

Maučec et al. found in their studies that TiO_2 and ZnO nanoparticles, with an average size of 5–8 nm, grafted onto ordered (SBA-15) and disordered (KIL-2) mesoporous silica, were evaluated against the model organic anionic dye Reactive Blue 19 (RB19). The nanoparticles broke down when exposed to UV light at pH = 7. The studies revealed that ZnO composites generally displayed stronger adsorption and quicker breakdown kinetics when grafted onto disordered KIL-2 silica, whereas opposite tendencies were seen for TiO_2.

6.3.3 *CARBON NANOPARTICLES*

The most often employed nano-adsorbents are carbon-based materials because of their wide availability, exceptional adsorption capabilities, strong chemical stability, and thermal stability. Studies show that the structure and surface charge of the dyes play important roles in the carbon nanotubes' ability to adsorb different dyes. For effective separation, several researchers have attempted to create a magnetic nanocomposite using carbon nanotubes as well. Other than carbon nanotubes, graphene-based nanomaterial is a viable option for the adsorption of inorganic and organic pollutants from aqueous solutions due to an improved surface-to-weight ratio and higher chemical stability.

Yang et al. reported that oxidized carbon nanoparticles (O-CNPs) prepared using the modified Hummers process showed good performance in adsorbing methylene blue [13]. The elimination rate was greater than 99.9% when the MB content was less than 50 mg/L. The absorbance curves were obtained and used to compare the MB solutions before and after the treatment.

6.3.4 GREEN SYNTHESIS

A significant method for removing dye molecules without producing additional hazardous chemicals is photocatalytic degradation of organic contaminants utilizing metal nanoparticles. Large quantities of metal oxides and pure metal nanoparticles have currently been successfully used to cause organic compounds to degrade. It is also noteworthy that bimetallic nanoparticles were effectively created by the straightforward adsorption approach [9]. Excellent catalytic activity for the reduction of nitrophenols and organic pollutants is shown by the produced Cu-Ag/GP and Cu-Ni/GP. Both Cu-Ag/GP and Cu-Ni/GP catalysts were used for the reduction of a mixture of 4-NP and RhB dye (Figure 6.2).

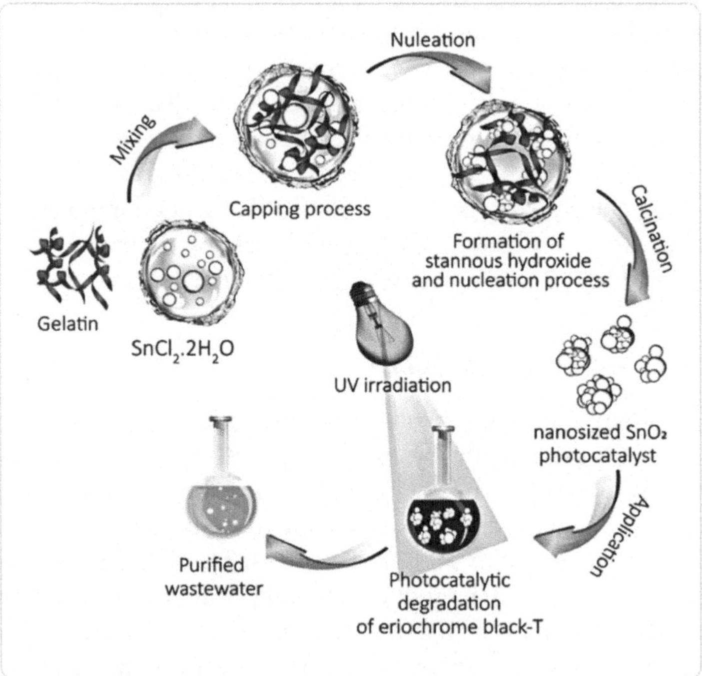

FIGURE 6.2 Tin oxide nanoparticles for water purification.

Source: Reprinted with permission from Ref. [15]. Copyright © 2020, Springer Nature B.V.

One of the effective methods for the removal of organic dyes from water includes photocatalytic degradation using metal nanoparticles as photocatalysts. Due to their large band gap and excellent energy stability, SnO_2 nanoparticles are recognized as remarkable photocatalysts. Morvarid Najjar et al. in their research developed an eco-friendly approach to produce SnO_2 nanoparticles in response to environmental concerns. The ideal approach for creating nanoparticles is to employ biocompatible natural polymers since they are abundant, non-toxic, economical, and environmentally friendly. In their work, a sol-gel method that uses gelatin as a complexing agent at various calcination temperatures is used to create SnO_2 nanoparticles in an environmentally friendly and simple manner. They synthesized SnO_2 nanoparticles that can act as effective photocatalysts for breaking down EBT dye [14].

6.4 CONCLUSION

Current advancements in nanotechnology provide a potential solution for water filtration that is affordable and superior in its capacity to remove impurities, and are reusable [5]. Now, as the globe struggles with a lack of clean water, researchers have demonstrated that nanoparticles, due to their unique properties, which include a large surface area, nanoscale size, extreme reactivity, high mechanical stability, strong solution mobility [14], dispersibility, hydrophilicity, and hydrophobicity, will be a better alternative for treating wastewater. The investigations of the various metal and metal-oxide based nanomaterials (zinc-oxide, titanium dioxide, nano-zerovalent iron, etc.) as well as carbon-based nanomaterials (carbon nanotubes, graphene, graphene oxide, fullerenes, etc.) and magnetic nanoparticles for the water treatment application clearly indicate the ability of nanomaterials in improving the environmental remediation system. The studies point to the fact that the most promising applications of nanotechnology in the environmental field have been in water treatment and remediation.

KEYWORDS

- **dye removal**
- **nanoparticles**
- **nanotechnology**
- **nano-zerovalent iron**
- **water pollution**
- **water treatment**

REFERENCES

1. Soha, M. A., Muhammad, I., Kalsoom, A., & Ekram, Y. D., (2019). Catalytic reduction of nitrophenols and dyes using silver nanoparticles at cellulose polymer paper for the resolution of waste water treatment challenges, *Colloids and Surfaces, A, 517*, 548–561.
2. Gholamreza, M., & Maryam, M., (2009). Removal of azo and anthraquinone reactive dyes from industrial wastewaters using MgO nanoparticles, *Journal of Hazardous Materials, 168*, 806–812.
3. Nkechi, P. N., Richard, M. M., Keneiloe, S., Oluwole, B. F., & Luqman, A. A., (2021). Antifouling polyethersulfone-petrol soot nanoparticles composite ultrafiltration membrane for dye removal in wastewater. *Membranes, 11*, 361.
4. Deepu, A. G., Daniel, P., Mariana, A. H., Luis, C. De. M., Yves, G., & Sabu, T. (2017). Meldrum's acid modified cellulose nanofiber-based polyvinylidene fluoride microfiltration membrane for dye water treatment and nanoparticle removal. American Chemical Society. *ACS Sustainable Chem. Eng.*
5. McMullan, G., Meehan, C., Conneely, A., Kirby, N., Robinson, T., Nigam, P., Banat, I. M., et al., (2001). Microbial decolorization and degradation of textile dyes. *Appl. Microbiol. Biotechnol., 56*, 81–87.
6. Opeyemi, A. O., Elias, E. E., Damian, C. O., & Maurice, S. O., (2020). Metal oxide-cellulose nanocomposites for the removal of toxic metals and dyes from wastewater. *International Journal of Biological Macromolecules, 164*, 2477–2496.
7. Yin, H., Qiu, P., Qian, Y., Kong, Z., Zheng, X., Tang, Z., & Guo, H., (2019). Textile wastewater treatment for water reuse: A case study. *Processes, 7*, 34.
8. Zhang, Y. R., Wang, S. Q., Shen, S. L., & Zhao, B. X., (2013). A novel water treatment magnetic nanomaterial for removal of anionic and cationic dyes under severe condition. *Chemical Engineering Journal*, S1385-8947(13)00916-9.
9. Muhammad, I., Khan, M. I., Sher, B. K., Murad, A. K., Kalsoom, A., & Abdullah, M. A. (2018). Green synthesis of plant supported Cu-Ag and Cu-Ni bimetallic nanoparticles in the reduction of nitrophenols and organic dyes for water treatment. *Journal of Molecular Liquids*, S0167-7322(17)35559-9.
10. Nasim, A., Mislav, M., Md Shahriar, A. H., Chaohai, W., Muxina, K., Jing, T., Jongbeom, N., Aslam, K., & Alan, R., (2020). Magnetic nanocellulose: A potential material for removal of dye from water. *Journal of Hazardous Materials, 394*, 122571.
11. Darja, M., Andraz, S., Alenka, R., Goran, D., Albin, P., & Natasa, N. T., (2017). Titania versus zinc oxide nanoparticles on mesoporous silica supports as photocatalysts for removal of dyes from wastewater at neutral pH. *Catalysis Today*.
12. Chang, J., & Waclawik, E. R., (2014). Colloidal semiconductor nanocrystals: Controlled synthesis and surface chemistry in organic media. *RSC Adv., 4*, 23505.
13. Sheng-Tao, Y., Jianbin, L., Jia-Hui, L., Qinghan, Z., Jing, W., Chen, M., Rong, L., et al., (2014). Carbon nanoparticles for cationic dye (methylene blue) removal from aqueous solution. *Nanoscience and Nanotechnology Letters, 4*, 839–842.
14. Haleema, S., & Syed, J. Z., (2020). Developments in the application of nanomaterials for water treatment and their impact on the environment. *Nanomaterials, 10,* 1764.
15. Morvarid, N., Hasan, A. H., Abdolhossein, M., & Alireza, H. M. D. (2020). Preparation of tin oxide (IV) nanoparticles by a green chemistry method and investigation of its role in the removal of organic dyes in water purification. *Research on Chemical Intermediates*.

16. Linic, S., Barteau, M. A., Graselli, R. K., & Burrington, J. D., (2008). *Handbook of Heterogenous Catalysis* (2nd edn., pp. 3458–3462). Wiley-VCH, Weinheim.
17. Zaccariello, G., Moretti, E., Storaro, L., Riello, P., Canton, P., Gombac, V., Montini, T., et al., (2014). TiO_2–mesoporous silica nanocomposites: Cooperative effect in the photocatalytic degradation of dyes and drugs. *RSC Adv., 4*, 37826.
18. Collard, X., El Hajj, M., Su, B. L., & Aprile, C., (2014). Synthesis of novel mesoporous ZnO/SiO_2 composites for the photodegradation of organic dyes. *Microporous Mesoporous Mater., 184*, 90–96.
19. Jeong, B., Kim, D. H., Park, E. J., Jeong, M. G., Kim, K. D., Seo, H. O., Kim, Y. D., & Uhm, S., (2014). ZnO shell on mesoporous silica by atomic layer deposition: Removal of organic dye in water by an adsorbent and its photocatalytic regeneration. *Appl. Surf. Sci., 307*, 468–474.
20. Zhai, J., Tao, X., Pu, Y., Zeng, X. F., & Chen, J. F., (2010). Core/shell structured ZnO/SiO_2 nanoparticles: Preparation, characterization and photocatalytic property. *Appl. Surf. Sci., 257*, 393–397.
21. Gnanaprakasam, A., Sivakumar, V. M., Sivayogavalli, P. L., & Thirumarimurugan, M., (2015). Characterization of TiO_2 and ZnO nanoparticles and their applications in photocatalytic degradation of azodyes. *Ecotoxicol. Environ. Saf., 121*, 121–125.
22. Rego, E., Marto, J., Marcos, P. S., & Labrincha, J. A., (2009). Decolouration of orange II solutions by TiO_2 and ZnO active layers screen-printed on ceramic tiles under sunlight irradiation. *Appl. Catal. A., 355*, 109–114.
23. Hevira, L., Zilfa, Rahmayeni, Ighalo, J. O., & Zein, R., (2020). Biosorption of indigo carmine from aqueous solution by *Terminalia catappa* shell. *J. Environ. Chem. Eng., 8*(5), 104290.
24. Ma, H., Burger, C., Hsiao, B. S., & Chu, B., (2011). Ultra-fine cellulose nanofibers: New nano-scale materials for water purification. *J. Mater. Chem., 21*(21), 7507–7510.
25. Brame, J., Li, Q., & Alvarez, P. J. J., (2011). Nanotechnology enabled water treatment and reuse: Emerging opportunities and challenges for developing countries. *Trends in Food Sci. Technol., 22*, 618.
26. Kovtyukhova, N. I., Ollivier, P. J., Martin, B. R., Mallouk, T. E., Chizhik, S. A., Buzaneva, E. V., & Gorchinskiy, A. D., (1999). *Chem. Mater., 11*, 771.
27. Choudri, B. S., & Baawain, M., (2015). Textiles. *Water Environ. Res., 87*, 1392–1413.
28. Bafana, A., Devi, S. S., & Chakrabarti, T., (2011). Azo dyes: Past, present and the future, *Environ. Rev., 19*, 350–371.
29. Khataee, A. R., & Kasiri, M. B., (2010). Photocatalytic degradation of organic dyes in the presence of nanostructured titanium dioxide: Influence of the chemical structure of dyes. *J. Mol. Catal. A Chem.*, 328–339.
30. Lee, K. M., Lai, C. W., Ngai, K. S., & Juan, J. C., (2016). Recent developments of zinc oxide based photocatalyst in water treatment technology: A review. *Water Res., 88*, 428–448.
31. Padhi, B., (2012). Pollution due to synthetic dyes toxicity & carcinogenicity studies and remediation. *International Journal of Environmental Sciences, 3*, 940.
32. El Gaini, L., Lakraimi, M., Sebbar, E., Meghea, A., & Bakasse, M., (2009). Removal of indigo carmine dye from water to Mg–Al–CO_3-calcined layered double hydroxides. *Journal of Hazardous Materials, 161*, 627–632.

33. Ayed, L., Mahdhi, A., Cheref, A., & Bakhrouf, A., (2011). Decolorization and degradation of azo dye methyl red by an isolated *Sphingomonas paucimobilis*: Biotoxicity and metabolites characterization. *Desalination, 274*, 272–277.
34. Haider, S., Kamal, T., Khan, S. B., Omer, M., Haider, A., Khan, F. U., & Asiri, A. M., (2016). Natural polymers supported copper nanoparticles for pollutants degradation. *Applied Surface Science, 387*, 1154–1161.
35. Gul, S., Rehan, Z. A., Khan, S. A., Akhtar, K., Khan, M. A., Khan, M. I., Rashid, M. I., et al., (2017). Antibacterial PES-CA-Ag2O nanocomposite supported Cu nanoparticles membrane toward ultrafiltration, BSA rejection and reduction of nitrophenol. *Journal of Molecular Liquids, 230,* 616–624.
36. Khan, S. A., Khan, S. B., & Asiri, A. M., (2016). Toward the design of Zn–Al and Zn–Cr LDH wrapped in activated carbon for the solar assisted de-coloration of organic dyes. *RSC Advances, 6*, 83196–83208.
37. Belessi, V., Romanos, G., Boukos, N., Lambropoulou, D., & Trapalis, C., (2009). Removal of reactive red 195 from aqueous solutions by adsorption on the surface of TiO_2 nanoparticles. *Journal of Hazardous Materials, 170*, 836–844.
38. Sohni, S., Gul, K., Ahmad, F., Ahmad, I., Khan, A., Khan, N., & Bahadar, K. S., (2017). Highly efficient removal of acid red and bromophenol blue dyes from industrial wastewater using graphene oxide functionalized magnetic chitosan composite. *Polymer Composites.*
39. León, G., García, F., Miguel, B., & Bayo, J., (2016). Equilibrium, kinetic and thermodynamic studies of methyl orange removal by adsorption onto granular activated carbon. *Desalination and Water Treatment, 57,* 17104–17117.

CHAPTER 7

A Case Study on Urban Stormwater Management for Sustainable and Resilient Measures: A Dutch Model

ASWATHY VIJAYAN,[1] S. S. PRAVEEN,[1] and S. SAJITH BABU[2]

[1]*Department of Physics, St. Thomas College, Kozhencherry, Pathanamthitta, Kerala, India*

[2]*Assistant Professor, Department of Physics, Catholicate College, Pathanamthitta, Kerala, India*

ABSTRACT

Netherlands is very much infamous for several disasters including floods. As part of their resilience methods, they adopted several different strategies and methodologies for reducing the impacts of floods. The state of Kerala was in the shadow zone of several natural disasters. But a series of natural disasters like Tsunami, Cyclone, and floods exposed the vulnerability of the state. The research and development associated with these disasters gained momentum and is still ongoing. This chapter reports different mechanisms implemented for flood protection in Dutch along with studies on implementing those in the state of Kerala for sustainable storm water management.

7.1 INTRODUCTION

Flooding is an unavoidable problem for the cities of Asia and Europe. Floods are the most common and worldwide naturally occurring phenomenon. In Kerala, also during the period of 2018 floods, 13 out of 14 districts were

Sustainable Environmental Engineering: Water Security, Energy Conservation, and Green Process.
Sonia Khanna, Ann Rose Abraham, Kian Hariri Asli, & A. K. Haghi (Eds.)

inundated [1]; Vishnu et al., 2019). Various countries across the globe also get inundated every year due to rainfall, tidal reflections, etc., in which the Netherlands shows some similar features related to Kerala. So Kerala and the Netherlands are at the forefront of the flood threat. The Netherlands adopted various safety measures to minimize and mitigate the flood and its impact. Even after replicated floods, the state of Kerala is yet to implement a flood warning system. The Netherlands is a country which is similar in size to the Kerala state. The Netherlands is the only country in the world that combines land use, environmental protection, and water resource management in the event of flooding due to climate change and rising sea levels. Studies say that unscientific construction activities and land use lead to natural disasters and increase the impact of disasters. Our state Kerala was a disaster-free state in India. But nowadays, our state is vulnerable to disasters like Tsunami (2004); Ockhi (2007); floods (2018, 2019, 2020, and 2021) [4, 5]. These disasters were really an eye-opener and a major shock to the people living in Kerala [2]. The impact of the floods of 2018 was very massive and tremendous. The frequently occurring floods exposed the reality that the state of Kerala is also vulnerable to floods [4]. The floods of 2018 affected the people and caused massive destruction to the properties, agricultural fields, water supply systems, transportation networks, and electricity supply, affecting around 54 lakh people. Among these, 14 lakh people were evacuated, and 433 people lost their lives. This chapter investigates the feasibility of adapting the Dutch model for Kerala as a protective mechanism to reduce the impacts of floods. The study also outlines measures for sustainable stormwater management through different methodologies.

The repeated disasters brought the Netherlands to the forefront of innovative and creative solutions in water management. This chapter is an investigation into the various dimensions of protective measures implemented by the research and development team of the Netherlands.

7.2 SIMILARITY BETWEEN KERALA AND NETHERLANDS

Kerala is known as the Gateway of the Monsoon. The state of Kerala is surrounded by the Arabian Sea on the west, the Western Ghats on the east, Karnataka on the north, and Tamil Nadu on the south. The state of Kerala covers an area of 38,863 km^2 [1]. The place Kuttanadu in the southern Alappuzha district is known as the Netherlands of Kerala. Kerala has 44 rivers, and out of those, 3 flow towards the east and the remaining flow towards the west. Kuttanadu is a region residing below sea level. The Netherlands is a country located in Europe. The word Netherlands means a low-lying country and covers an area of 41,526 km^2. In the Netherlands, 26% of the area is lying below sea level, and the remaining 60% is flood-prone areas. The Netherlands has low elevation and flat topography, and it is classified as Low Netherlands (LN) and High Netherlands (HN) based on topography. Low land contains many artificial and natural waterways. Eight major rivers are flowing through the Netherlands.

7.3 FLOODS IN KERALA

The state of Kerala, located in the southern peninsula of India and residing on the west coast of India, was in the shadow zone of several different natural disasters. But various disasters that occurred between 2004 and 2018 exposed the vulnerability of Kerala state. Prior to that, Kerala experienced several floods in the years 1907, 1924, and 1960. The floods of 1907 occurred between July and August, during which Kerala received 1,780 mm of rainfall, 175% higher than the normal rainfall in Kerala [3].

The floods of 1924, which happened in July, are regarded as one of the worst floods. The unexpected opening of many dams also resulted in this flood. Kerala received 3,368 mm of rainfall. The floods of 1960 occurred from June 4 to August 21. Kerala received 2,387 mm of rainfall, which was 50% higher than the normal rainfall. The floods of 2018 occurred in the month of August. Improper dam management, improper land cover, and land use activities are regarded as the main causes of the 2018 flood. Kerala received 23,466 mm of rainfall between June 1 and August 20. Seven districts experienced severe flooding. The flood of 2019 was comparatively less severe than the 2018 flood. A total of 121 people died in the 2019 flood, which happened between August 8 and August 29. The floods of 2018 were an eye-opener to the people of Kerala because this was the most devastating one that occurred during a recent time period [3].

7.4 FLOODS IN NETHERLANDS

Some historically recorded flood events which happened in Netherlands are shown in Table 7.1.

TABLE 7.1 Historical Floods in Netherlands

Sl. No.	Year	Name of Flood	Loss and Death
1.	1287	St. Lucia's flood	50,000 to 80,000
2.	1421	St. Elizabeth's flood	2,000 to 10,000
3.	1530	St. Felix's flood	1,00,000
4.	1916	Zuiderzee flood	–

7.4.1 ST. LUCIA'S FLOOD

It was a storm tide which happened on 13th and 14th December 1287, and this storm is referred to as the Great Storm. St. Lucia's flood affected prominent parts of the Netherlands and Northern Germany. Due to this flood, some areas were completely submerged. The flood created a death toll of approximately 50,000 to 80,000 people and is regarded as one of the largest floods in world history.

7.4.2 ST. ELIZABETH'S FLOOD

During the night of 18th and 19th November 1421, a heavy storm occurred nearer to the North Sea coast, which caused floods and highly affected the dikes. During this flood, the dikes in many places collapsed, and the lower-lying polder landscape areas also got flooded, submerging many villages and killing around 2,000 to 10,000 people.

7.4.3 ST. FELIX'S FLOOD

St. Felix's flood happened on 5th November 1530. More than 1,00,000 people were killed in the Netherlands by the St. Felix's flood. Large parts of Flanders and Zeeland were washed away in this flood.

7.4.4 FLOOD OF 1916 OR ZUIDERZEE

This flood happened in the night between 13 and 14 January 1916 in the Netherlands along the dikes of the Zuiderzee as a result of a storm surge. It destroyed many villages.

7.5 PROTECTIVE MECHANISMS

Replicated floods evoked a sense of fear for the Dutch residents and exposed the need for initiating research and development for framing and implementing protective measures. The people of Dutch protected their country by building dams, streams, and canals. Polder and dike are the major mechanisms that are used to conserve the land use and water management in the Netherlands.

7.5.1 POLDER

The Netherlands is the first country to introduce the concept of poldering. The process of creating land from seas, lakes, and marshy areas is called poldering. The land which is created by restricting the entry of sea and river water with the help of protecting walls is called a polder (Figure 7.1). A polder is a combination consisting of land and water areas which are completely enclosed by dikes. The main aim of this construction is to drain the wetlands and make it suitable for agricultural and other construction activities. If the land surface is above the low-tide level, the water may be drained out to the sea through tide gates. It prevents the intrusion of seawater at the time of high tide or tidal reflection. If the land is lying below low-tide level, the water is pumped through the dikes. The large accumulation of sediments in the low-lying areas helps to build up the polder from the bottom to a higher level, which is suitable for agriculture.

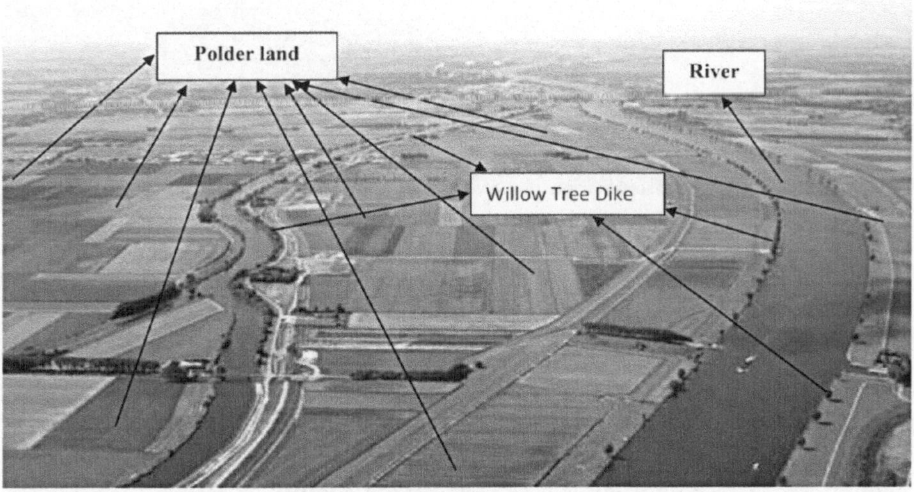

FIGURE 7.1 Thesilted-up land from river (polder).

About half of the country's land, 18,000 sq. km, or 50% of the land of the Netherlands, was converted into polders as a protective measure.

7.5.2 DIKE

A dike is a protective wall used to protect the polder and other regions that lie below sea level, as depicted in Figure 7.2. A dike is a flood-preventing barrier that helps low-lying areas overcome the threat of floods. It prevents the intrusion of sea and river water. A dike is a permanent structure made of rock and is mainly constructed to protect areas lying below the mean sea level, aimed at reducing erosion, surface runoff, and wave action.

FIGURE 7.2 The protection dike barrier in Netherlands.

7.5.3 POLDER LANDSCAPE

Polder landscape is the land area that has been reclaimed from the sea, rivers, and peat by restricting the entry of river water and then allowing the flow of sea water from this polder to the outside terrain. The region is protected by using dikes. The polder landscape is shown in Figure 7.3.

FIGURE 7.3 The sky view of polder landscape.

Today, land use, water management, and environmental protections are fully under the control of the water board governed by local bodies. For water management and flood control, the Netherlands has established a national water authority called Rijkswaterstaat (RWS). The main function of Rijkswaterstaat (RWS) is to manage the main highways and the national water system in the Netherlands. It also ensures a sufficient supply of clean water in the Netherlands. RWS primarily works to ensure that traffic can flow smoothly and safely along more than 6,000 km of highways and waterways without causing any problems.

7.6 IJSSELMEER AND AFSLUITDIJK

After the Zuiderzee flood in 1916, a decision was made to reclaim the sea in 1927, leading to the creation of a shallow freshwater lake named Ijsselmeer. This artificial lake was developed by constructing dikes. Ijsselmeer is a freshwater lake in the Netherlands and it acts as a massive freshwater reservoir, as shown in Figure 7.4. It ensures the availability of freshwater in cities and for agricultural purposes. It encloses 11,000 km with a depth of 5.5 m. Water from the Rhine flows into the Ijsselmeer, which in turn acts as a freshwater lake. The Ijsselmeer is mainly used for transportation purposes and fishing, providing a livelihood for a huge population. It offers numerous opportunities for people, both on the water and on its shores.

FIGURE 7.4 Freshwater reservoir in Netherlands.

The Afsluitdijk is a major dam and causeway in the Netherlands constructed after filling Zuiderzee as shown in Figure 7.5. It was constructed between 1927 and 1932. The inland side is made of heavy stone. The seaward side is made of boulder clay with willow wood mattresses. The bottom of the Afsluitdijk is built using basalt boulders or rocks and old concrete. The structure of the Afsluitdijk is such that boulder clay is superior to just sand or clay. This could be retrieved by dredging it from the bottom of the Zuiderzee. Barite is a mineral which is also used as a construction material to save the dikes from corrosion. These form a level block. It has a strong wave retardant effect because each dike block weighs 6,500 kg. The outer side of this dike consists of many claddings and it also helps to protect the dike from heavy storms and waves. This is overflow-resistant. Afsluitdijk regulates the water level of the Ijsselmeer and provides protection for the Netherlands from the Wadden Sea.

7.7 MAESLANTKERING

The Maeslantkering is the largest movable storm surge barrier in the world. It consists of two massive steel arms and has the ability to block a storm surge of up to five meters (Figures 7.6 and 7.7). It was constructed from 1991 to 1997. The Maeslantkering barrier has two 210-meter-long barrier gates and a height of 22 meters. The barrier gates are made up of iron pipes

FIGURE 7.5 The national highway system on Afsluitdijk.

with ball joints of 10-meter diameter and a weight of 680 tons. The barrier is fixed on either side of the canal with a depth of 15 meters. The barrier is connected to a computer system which is linked to weather and sea level data. The computer will monitor the sea level every 10 minutes during the storm season. This system is easily programmed to shut the barriers when the water level rises to 3 meters, so it can be automatically closed when Rotterdam is threatened by floods due to the worst weather or rising water level. When the sea water level rises, the water flows into the canal and fills the connected pipes. After that the water-filled iron pipes will be completely submerged into the water, and it will become a strong barrier.

It contains 30 pumps to pump out water in 90 minutes and the arms return to their normal position. The main aim of this barrier is to protect the Rotterdam Harbor from flooding.

7.8 FLOOD PREVENTION METHODS IMPLEMENTED IN THE NETHERLANDS – A BRIEF OVERVIEW

The Netherlands used three methods to mitigate or prevent floods. They are Building with Nature, Living with Water, and Room for the River.

FIGURE 7.6 The overview of storm surge barrier.

FIGURE 7.7 Imagery of closed Maeslantkering.

7.8.1 BUILDING WITH NATURE

Approximately 350 km of coastal areas are prone to flooding. This is mainly due to the reduction of delta deposition, depression of land area, and the increasing water level of the sea. In Zuid Holland, every five years, the coastal areas are strengthened by dumping soil. However, this process damages the ecosystem and fertility of the soil. Building with nature includes the zand motor, oyster reef, and willow tree dikes.

7.8.1.1 ZAND MOTOR

Zand motor is an artificial sandbank. Large-scale deposition of sand is converted into sand dunes with the help of wind and waves, which is called zand motor. The construction process was completed in November 2011. The main aim of this artificial sandbank is to reinforce the coastline for a long period of time, providing long-term safety for the coastline. It also helps to create an attractive area for people to enjoy the beauty of nature. This is a resistance against sea level rise and it prevents coastal flooding. The time span required for creating this mechanism is approximately 8 months, and an amount of 200 cubic meters of sand is used for creating this zand motor, as shown in Figure 7.8. It looks like a hook-shaped peninsula that extends up to 1 km in length and 2 km in width. It acts as a buffer against sea level rise and helps to reduce the effects of storm surge and coastal flooding.

FIGURE 7.8 The hook shaped Zand motor.

7.8.1.2 OYSTER REEF

Oysters are the most valuable assets of the ocean. Commonly, oysters are grown on salty and submerged surfaces, piers, and rocks. The highly strengthened tidal waves flap on the coastal areas when the tidal barriers get destroyed. To overcome this problem, the authorities made tidal barriers by allowing the oyster larvae to grow on a hard substrate such as rocks or iron.

The accumulation of oysters forms a huge oyster colonial community. After that, new oyster reefs may form on the substrate, and it is called an oyster reef. This helps to accumulate delta soil in the river mouth. It also provides habitat for a large number of marine animal and plant species. A single oyster has the ability to filter around 50 gallons of water per day. It also reduces the effects of waves, floods, and tides. Therefore, oyster reefs function as a natural filter. The pattern of the oyster reef is shown in Figure 7.9. It was successfully established in Eastern Skeld in 2010.

FIGURE 7.9 Oyster reef barrier.

7.8.1.3 WILLOW TREE DIKE

Instead of building dikes by using rocks with concrete, Netherlands scientifically grows willow trees in front of the protection barriers. The willow tree acts as a barrier by attenuating the wave energy. The wave is encountered by most of the tree surface. The tree leaves are responsible for attenuating the wave energy. It helps to prevent around 80% of the waves from rising. The preventive method using willow trees is shown in Figure 7.10.

FIGURE 7.10 Willow tree dikes in Netherlands.

7.8.2 LIVING WITH WATER

Living with water is a concept that the Netherlands adopted to mitigate the impact and loss caused by floods.

7.8.2.1 FLOATING HOUSES

In the Netherlands, Masbomel has built 42 floating houses, known as amphibious houses, as shown in Figure 7.11. These houses are constructed on wood and fixed on iron pillars. The basements of these houses are made using hollow concrete square blocks. As the water rises, the houses will be raised on iron pillars. This type of construction ensures the proper functioning of floating homes.

7.8.2.2 WATER SQUARE

The water square in Tiel is mainly aimed at flood control and creating a habitable environment. During the summer seasons, this place is used as a sports and entertainment area, coordinated by the water board and the community. However, during the monsoon season, it functions as a water collection unit

FIGURE 7.11 Floating houses in Maasbommel.

in Tiel, capable of holding up to 550 cubic meters of water. This construction is primarily intended to protect the low-lying areas from the water flow of the Vall river. The design of the water square is illustrated in Figure 7.12.

FIGURE 7.12 Water square in Rotterdam city.

7.8.2.3 FLOOD BRIGADE AND KEI ARMY

The Flood Brigade and KEI Army are dynamic groups of people assigned with the task of collecting and storing the necessary materials required for the construction of temporary protective walls in a safe place. The KEI Army is also involved in building protective walls using sandbags in the event of flooding.

7.8.2.4 PARKING GARAGE

One of the most amazing things in Rotterdam city is the underground parking garage known as Museum Park. It is used as a water storage system during the monsoon season, collecting 10 billion liters of water every 30 minutes. After the monsoon season, the water in the storage is allowed to flow out from this garage. Kruispleain is another underground parking garage similar to the Museum Park, accommodating 760 vehicles. It has the ability to collect 2.4 billion liters of water and is connected to the neighboring canal networks. This water can be purified and used for various purposes, including irrigation.

7.8.3 ROOM FOR THE RIVER

The "Room for the River" concept is an artificial mechanism implemented in a natural way to control floods and their impacts. This is achieved by increasing the height of dikes and widening the river, facilitating the smooth flow of water. This method is highly regarded as the best way to mitigate the effects of flooding due to varying climatic conditions.

The "Room for River" method can be implemented as follows:

- Allowing the smooth flow of water by avoiding the sedimentation of delta soil and slopes situated in flood-prone areas.
- Increasing the depth of the river by removing the soil and clay at the bottom, contributing to the ensuring flow of the river by lowering the floodplains.
- Constructing parallel protection walls by replacing obstructing walls, thereby allowing excess water to drain away.
- Strengthening the retaining wall at a point where the river area can't be increased.

- Relocating the retaining wall towards the shore will help to increase the area of the riverbed.
- Instead of digging the bottom to drain the excess water, alternative ditches are made by building retaining walls on both sides of the river.
- The sea walls and docks should be lowered.

The Netherlands implemented the "Room for the River" project in Tevendercity. They constructed new dams, cattle farms, and grazing fields beside the river as a part of the "Room for the River" project.

7.9 KERALA'S ADAPTATION

Netherlands is one of the most densely populated countries in the world. It is the only country that uses sophisticated methods to prevent flooding by incorporating water management, land use, and land cover. The Netherlands is most renowned for its flood control measures and methods. The repeated and continuous struggle with water has led them to invent and adopt these types of prevention methods. They have constructed dikes, dams, canal systems, and flood gates to protect their land from the sea, rivers, and floods. Netherlands has experienced some of the most devastating floods and storm surges in world history, but they have survived through their highly innovative ideas. The state of Kerala is similar to the Netherlands in terms of land area, topographical pattern, terrain undulations, and geological features. Since 2004, the state of Kerala has also been vulnerable to many disasters, including floods. Therefore, adopting and strictly following the above-mentioned Dutch-initiated flood prevention measures to mitigate the impacts of floods would be commendable. Building with nature and Room for the River can be regarded as effective mechanisms to be implemented in Kerala. Similarly, several methods can be adopted and implemented in different ways. The willow tree dikes in the Netherlands can be substituted by mangrove forests along the coastal areas in Kerala. They can act as a fence and also provide habitat for several marine plants and animals. This will also help in minimizing the effects of coastal erosion, tides, and back-water reflection. The technique of the Sand Motor can also be applied in Kerala scenarios, which could prove beneficial to the fishing community in Kerala. The concept of "room for the river" is a necessity in the Kuttanadu region as the polder region in Kuttanadu is under threat of flooding. Adaptive measures like living with water are not uncommon in Kuttanad. Especially

Thanneermukkam bunds, built on poles, are not new in this region. Increased urbanization has resulted in massive dredging of sand from rivers, sand filling of ponds, wetlands, and water bodies. However, these measures are not capable of mitigating the impacts of floods. Therefore, different stakeholders associated with flood modeling, mapping, and mitigation should try to adapt the greatly acknowledged Dutch model in Kerala as part of the "Rebuild Kerala" initiative to reduce flood vulnerability.

7.10 CONCLUSION

This chapter discusses various flood protection mechanisms adopted in the Netherlands. Kerala and the Netherlands share similar topography and experience similar patterns of flooding. The different protective measures, such as building with nature, oyster reef experiments, sapling plantation techniques, and creating room for the rivers, can be successfully implemented in the state of Kerala to provide protection from disasters. This chapter recommends that different stakeholders involved in modeling and mapping should adopt these types of mechanisms to reduce vulnerability.

ACKNOWLEDGMENTS

The authors are thankful to the management, principal, and staff of St. Thomas College Kozhencherry. The authors are indebted to the faculty members of the Department of Physics, St. Thomas College Kozhencherry, for their support and understanding.

KEYWORDS

- **dike**
- **flood protection mechanisms**
- **floods**
- **plantation techniques**
- **polder**
- **sustainable storm water management**
- **vulnerability**

REFERENCES

1. Sudheer, K. P., Bhallamudi, S. M., Narasimhan, B., Thomas, J., Bindhu, V. M., Vema, V., & Kurian, C., (2019). Role of dams on the floods of August 2018 in Periyar River Basin, Kerala. *Current Science (00113891), 116*(5).
2. Jasmine, P. R., Susan, M., Praveen, S. S., & Sajith, B. S., (2020). *Estimation of Clear Characteristics Using Numerical Simulation Along Chengannur to Kozhencherry, Stretch on The Banks of Pampa River, Southern Kerala.*
3. Srija, P., Nithin, G., Chaithanya, V., Sushma, S. M., & Sridevi, M., (2021). A case study on Kerala floods. *International Journal of Current Engineering and Technology,* 26–31.
4. Praveen, S. S., (2018). Numerical modeling of floods in Kerala-a research and developmental perspective. *Rational Discourse, 24*(1).
5. Praveen, S. S., Veena, S. B., Aswathy, V., & Sajith, B., (2018). *Numerical Modeling of Flood Inundation and Run Up Along the Banks of Pampa River, Southwest Coast of India.*
6. Vishnu, C. L., Sajinkumar, K. S., Oommen, T., Coffman, R. A., Thrivikramji, K. P., Rani, V. R., & Keerthy, S., (2019). Satellite-based assessment of the August 2018 flood in parts of Kerala, India. *Geomatics, Natural Hazards, and Risk, 10*(1), 758–767.

Index

D

E

F

P

Q

R

S

U

V

W

Y

Z